Industrial Productivity and International Technical Cooperation

The Technology Policy and Economic Growth Series

Herbert I. Fusfeld and Richard R. Nelson, Editors

Fusfeld/Haklisch INDUSTRIAL PRODUCTIVITY AND
INTERNATIONAL TECHNICAL COOPERATION
Fusfeld/Langlois UNDERSTANDING R & D PRODUCTIVITY
Hazewindus U.S. MICROELECTRONICS: Technical Base,
Industry Structure, Social Impact
Nelson GOVERNMENT AND TECHNICAL PROGRESS:
A Cross Industry Analysis

Pergamon Titles of Related Interest

Dewar INDUSTRY VITALIZATION: Toward a National
Industrial Policy
Hill/Utterback TECHNOLOGICAL INNOVATION FOR A
DYNAMIC ECONOMY
Lundstedt/Colglazier MANAGING INNOVATION: The Social
Dimensions of Creativity
Perlmutter/Sagafi-nejad INTERNATIONAL TECHNOLOGY
TRANSFER
Sagafi-nejad/Moxon/Perlmutter CONTROLLING INTERNATIONAL
TECHNOLOGY TRANSFER
Sagafi-nejad/Belfield TRANSNATIONAL CORPORATIONS,
TECHNOLOGY TRANSFER AND DEVELOPMENT

Related Journal*

Technology in Society

*Free specimen copy available upon request.

Industrial Productivity and International Technical Cooperation

edited by
Herbert I. Fusfeld
Carmela S. Haklisch

The Technology Policy and Economic Growth Series,
Herbert I. Fusfeld and Richard N. Nelson, Editors

Published in cooperation with the Center for Science and Technology
Policy, Graduate School of Business Administration, New York University.

Pergamon Press
New York Oxford Toronto Sydney Paris Frankfurt

Pergamon Press Offices:

U.S.A.   Pergamon Press Inc., Maxwell House, Fairview Park,
                    Elmsford, New York 10523, U.S.A.

U.K.   Pergamon Press Ltd., Headington Hill Hall,
                    Oxford OX3 OBW, England

CANADA   Pergamon Press Canada Ltd., Suite 104, 150 Consumers Road,
                    Willowdale, Ontario M2J 1P9, Canada

AUSTRALIA   Pergamon Press (Aust.) Pty. Ltd., P.O. Box 544,
                    Potts Point, NSW 2011, Australia

FRANCE   Pergamon Press SARL, 24 rue des Ecoles,
                    75240 Paris, Cedex 05, France

**FEDERAL REPUBLIC Pergamon Press GmbH, Hammerweg 6
OF GERMANY**   6242 Kronberg/Taunus, Federal Republic of Germany

Library of Congress Cataloging in Publication Data
Main entry under title:

Industrial productivity and international
    technical cooperation.

"Based on presentations from a conference
entitled Industrial Productivity and International
Technical Cooperation held in Paris, France,
on Nov. 20-21, 1980, by the Center for Science and
Technology Policy, Graduate School of Public
Administration, New York University."
    1. Technology--International cooperation--
Congresses. 2. Industrial productivity--Congresses.
I. Fusfeld, Herbert I. II. Haklisch, Carmela S.
T49.5.153   1982   338.91   81-19189
ISBN 0-08-028810-3   AACR2

Printed in the United States of America

Contents

v

Preface

Herbert I. Fusfeld
Carmela S. Haklisch

The papers and summaries of discussions contained in this volume are drawn from a conference on industrial productivity and international technical cooperation, held in Paris on November 20 and 21, 1980, with participants from the United States, Europe, and Japan. To the best of our knowledge, this was the first time that the attention of experienced senior representatives from both industry and government was directed collectively on the relationship between the private sector and cooperative technical agreements among governments. The overall objective of the conference was to examine opportunities for multilateral cooperation involving the private sector that might enhance productivity of the research and development process and thus of industrial productivity. The conference therefore focused on: (1) the feasibility of greater interaction between the industrial research community and the activities of international cooperative agreements among governments in science and technology, and (2) the development of guidelines for future agreements that could improve their effectiveness in terms of technical priorities, transfer of technology, and application of results.

The program of the conference was not designed to be exhaustive or philosophical.' Rather, it was designed to be an initial, realistic examination of an existing situation, of potential interactions, and of possible improvements. The chapters of this volume are based on the major papers from the conference program, including the formal papers presented at the plenary sessions, the papers presented at each panel session to stimulate discussion, and summaries of each panel discussion.

Perhaps the greatest contribution which this conference made was to increase awareness within the industrial research community of the span of activities conducted under international technical agreements. Far from being complete, this volume represents initial insights that might advance further study and thus serves as a valuable introduction to the subject. Hopefully, this work and the work to follow may result in a broader range of substantive options, recommendations, and improvements for the international technical community.

The conference represented a truly international enterprise. We are deeply grateful to François Davoine (professor) and Francis Cambou (director) of the Conservatoire National des Arts et Métiers for their boundless assistance and gracious hospitality in providing facilities for the meeting. The French government generously arranged for simultaneous translations of the plenary sessions at the conference. Dr. Abraham Friedman, science counselor for the American embassy in Paris, was a continuing source of logistical support and Bernard M.J. Delapalme, Directeur de la Recherche Scientifique, of ELF Aquitaine, France served as a conference session chairman.

We also wish to express our deep gratitude to the Division of International Programs (INT) of the National Science Foundation of the United States for providing a major portion of the funds to support this effort.

INT program activities include bilateral cooperation in science between the U.S. and other countries. Conference results are expected to provide guidelines to the Division on increasing industry participation in INT-sponsored relationships between the U.S. science community and its counterparts abroad. We particularly wish to thank Dr. Bodo Bartocha, Division Director, INT, for his continuing support of this project.

We tremendously appreciate the participation of several senior colleagues who represented government perspectives in a panel chaired by A.E. Pannenborg, Vice President, Board of Management, N.V. Philips Co. The panel included:

- Pierre Aigrain
  Secrétaire d'État de la Recherche, France
- Duncan Davies
  Chief Scientist and Engineer, Department of Industry
  United Kingdom
- Wolfgang Finke
  Ministerialdirektor, BMFT, Germany
- Hiroshi Inose
  Professor, Department of Electronic Engineering,
  University of Tokyo, Japan
- Francis Wolek
  Deputy Assistant Secretary for Productivity, Technology
  and Innovation, Department of Commerce, United States

Special thanks are particularly due our dedicated staff for their good humor and professionalism in handling the myriad details associated with the conference and the preparation of these proceedings. We wish to thank the participants of the conference for taking the time to engage in this effort and, most of all, the authors of the following chapters for their abundant cooperation throughout the publication process.

I
Overview

1

Changing Trends in International Technical Cooperation

Herbert I. Fusfeld
Carmela S. Haklisch

The increasing importance of science and technology to industrialized societies, and the growing interdependence of all nations, have combined to spawn a new growth activity, the proliferation of cooperative agreements in science and technology entered into by the governments of two or more countries. There are many categories of such agreements, and they are designed to serve many different objectives. Most are bilateral agreements between two countries. Many are multilateral agreements among a number of countries or within a geographical region. Many agreements emphasize basic scientific research. Others focus on specific technologies or mission-oriented goals.

A principal objective behind many agreements is to generate and disseminate a broader reservoir of basic science and engineering. In recent years, the rising costs of research have stimulated cooperation in order to share expenses. The nature of certain technological problems requires international cooperation for their solution, such as common flight-control equipment and the regional impact of air or water pollution. In many instances, the desire to promote good relations between countries has led to a cooperative technical agreement as a mechanism for developing interactions among research personnel, leading to appreciation for each other's capabilities and procedures, while conducting some constructive activity of mutual benefit.

There is one feature that is common to a great many of these agreements. They are initiated most often by government representatives who identify a need or an opportunity for technical cooperation as a mechanism for contributing to particular objectives of the countries involved. The planning, conduct, and dissemination of R&D that make up these agreements are then administered and usually staffed by government personnel. The conceptual basis for these agreements and the perspective on how the results are to be integrated within the economies of the countries involved are provided by public representatives attempting to aid objectives of a broad public nature.

In brief, as a general rule, there has been little involvement of private sector representatives in the initiation and implementation of the great bulk of these cooperative agreements among governments. There are, of course, technical representatives from industry invited to serve on advisory committees, and this does indeed serve to link an awareness of current industrial research to the activities of the cooperative agreements. Industry also performs research under contract to government agencies as part of the work of technical agreements. Nevertheless, the participation expected from these industry individuals relates primarily to their technical expertise and would not normally include judgments related to the introduction of technical change — judgments on investment, manufacturing, or marketing capabilities.

THE TIE TO PRODUCTIVITY

The unique function of private sector industrial research is to provide options for technical change which can be realistically integrated into the business plans and operations of the corporation and ultimately into the economy. This is the principal mechanism by which R&D contributes to industrial productivity. This, however, is not normally the purpose, function or the anticipated result of international technical cooperation.

The issue considered at the conference which gave rise to this volume was whether the industrial research efforts of the private sector can benefit from some type of interaction with international technical agreements, and thus make more effective the role of R&D in improving industrial productivity. There is the further issue of whether, once private sector cooperation is stimulated, such cooperation will contribute to the primary objectives of the agreement. The conference participants discussed the issue of private sector interests in different areas of basic science and in different areas of technology. They sought to relate potential benefits in each area to the mechanisms necessary for involving the private sector. The limitations of such cooperative agreements were also identified.

The question for the private sector was not whether there should be such international agreements; that is the responsibility of the governments involved, and is based on objectives that touch on economics and foreign policy as much as on science and technology. The question, rather, was whether the very considerable technical efforts that are being conducted under these agreements can be of more value to the private sector than at present, what public-private cooperation this might require, and how this could strengthen both sectors.

The objective of the conference was therefore to develop suggested guidelines for the design and implementation of international cooperative agreements among governments as seen from the perspective of the private sector. The justification for this effort is that such public-private interactions will expedite the introduction of technical change into the economies of the countries involved, and thus strengthen the objectives sought by the agreements. Put another way, if such international cooperative agreements are structured to attract the participation of the private sector, then this may ensure increased contributions to industrial productivity from a broader reservoir of science and technology and thus from a stronger base of industrial research, and will more likely achieve the goals of the public sector as well.

The objectives of private industrial research are not the same as those of governments. There are obvious limitations to what collective efforts of a public nature can do to improve the effectiveness of competitive private corporations in the use of such technical activities. But there are indeed opportunities in such areas as lowering costs, broadening the technical base, and in pursuing common interests related to health, safety, and energy conservation. A stronger, more effective industrial research community is the best mechanism for applying science and technology to the improvement of industrial productivity. This not only is a principal objective of every private corporation, but represents collectively perhaps the most important public objective of the industrialized societies. Since the private and public sectors have this broad goal in common, the conference sought to define what cooperation is possible in the research efforts of both sectors that can approach this goal most effectively.

INTERNATIONAL COOPERATIVE AGREEMENTS

The variety and scope of international cooperative agreements reflect different characteristics among many scientific and technical fields and many different objectives that cause these agreements to be initiated. The objectives include the following:

1. Cost Sharing. Certain investigations require such very expensive facilities (nuclear energy, space) or massive concentration of resources (synthetic fuels, ocean drilling) that interested participants can derive great economies through a joint research program focusing on technical problems of common interest.

2. Standardization. The nature of technology today and of international trade often require agreements among countries on technical specifications for products and systems. These may include telecommunications, transportation, air flight controls, and so on.

3. Strengthening Basic Science. This has been the traditional focus for international cooperation in areas where there are opportunities for joint research or exchanges of people and in which the results are of common interest and common usefulness to the cooperating countries. The intent is to expedite the rate of progress in a field and to stimulate diffusion of results. This presumably leads to cost-effectiveness and a greater reservoir of knowledge for the world scientific community.

4. Improving International Political and Economic Relations. The desire to strengthen political and economic ties is often the primary motivation for a technical agreement. In this instance, the commitment to some level of effort comes first, and the detailed subject matters are selected which are most likely to be useful and relevant to the needs of the individual countries involved.

5. Solving Particular International Problems. There are certain problems whose nature requires international cooperation, as in weather forecasting or environmental control of acid rain. Others are simply of such urgent common interest that cooperative agreements appear logical, as in energy conversion and conservation.

The variety of objectives underlying international technical cooperation has produced a very large and rapidly expanding level of effort. The subjects cover diverse fields of science and technology. The specific mechanisms used in the implementation of these agreements range from exchanges of information at conferences and seminars, to exchanges of personnel, to actual joint research efforts. The agreements are usually between two countries; often include a group of countries, as in energy and environmental concerns; and are sometimes regional, as in the case of the European communities.

The total amounts of money and technical personnel now engaged in these cooperative efforts are very substantial. Regardless of the objective being served by any particular agreement – technical, political, or economic – the sum of these activities constitutes an appreciable world-wide technical effort forming an important component in the development and use of technical advances.

The question, then, is not whether these agreements and activities should exist, since the many legitimate objectives and interests served by the agreements ensure that they will indeed continue and grow. Rather, the focus is on whether the private sector, as the principal instrument for integrating science and technology into society, can participate in and benefit from these agreements, and whether, in so doing, it can strengthen the objectives of the agreements themselves and strengthen the broad technical base of the industrialized countries.

COOPERATION IN SCIENTIFIC AND TECHNICAL RESEARCH

One particular program, European Cooperation in the Field of Scientific and Technical Research (COST), is an illustration of the contribution that technical cooperation can make. The system was established in 1970 as a mechanism to facilitate R&D cooperation among member states of the European Community and nonmember European countries. The impetus for launching this effort was based primarily on four objectives:

● To maintain the pace of technological development within the community by pooling resources and personnel

- To establish closer cooperation with potential members
- To extend the community's research activities to areas not addressed by formal treaties
- To establish concerted action projects as a new mechanism for international cooperation.

After an extensive review of over forty projects, seven were selected, each of which involved a different group of countries. Each project operated under a separate international agreement, and each partner country paid its proportionate share of the project – an arrangement dubbed "a la carte" financing.

One project underway since 1977 involves collaborative research in redundancy reduction techniques for visual telephone signals, with participating laboratotires in Belgium, France, the Netherlands, Italy, Sweden, West Germany, and the United Kingdom. Another project is devoted to optical fiber communication systems. Other projects include studies on the properties and behaviors of materials used in gas turbines and studies to identify and analyze the presence and concentration of organic pollutants in air and water.

Private sector involvement in the COST program varies according to country participation and the needs of each project; it may include the sale of equipment and material, involvement as a subcontractor to a government laboratory, or direct participation in research or testing. Provisions are made for communicating technical data, and arrangements are worked out to accommodate patent and licensing interests and rights of industrial property.

CONFERENCE PERSPECTIVES

The COST program is but one of the numerous models that exist for international technical cooperation. The chapters in this volume describe other examples and provide insights into the interests and participation of industry in the activities of international technical agreements. To offer perspective for considering this subject, M. Jacques Desazars de Montgailhard poses and comments on three questions: How is industrial productivity enhanced? What is the attitude of industrial concerns to cooperation? How can technical cooperation be expanded in different types of industrial activity?

Robert Hawkins provides an economic explanation of the costs and benefits tied to cooperation and the impact on world productivity growth. Klaus-Heinrich Standke details for reference purposes the network and modus operandi of international technical agreements. J.E. Goldman focuses on international mechanisms for cooperative arrangements between companies that might provide an improved use of resources.

Four fields in science (materials, chemistry, biology, and electronics) and four fields in technology (energy conversion and conservation, transportation, telecommunications, and environmental control) were examined to consider issues relevant to industrial participation within the context of a specific technical area. This review produced two categories of information. The first includes issues related to technical growth for each field: What are the principal subfields receiving the greatest effort in R&D today, or holding great potential interest? Among these, which subfields or areas offer the most promising opportunity for international cooperation? What are the areas of research where competitive interests prevent any participation at all in international agreements?

Due to the diverse nature of the eight fields reviewed and the competitive considerations related to critical subfields, a different tapestry emerged for each area, with suggestions and models for study. For example, due to the intense competition in the electronics field, one might not immediately consider frontier research in this area as a possibility for collaboration. However, Michiyuki Uenohara describes a project undertaken in Japan called the Cooperative VLSI (Very Large Scale Integration) Research Project. In this program, competing Japanese computer manufacturers worked together in a basic portion of VLSI technology. Each company participating pursued its own development efforts on a proprietary basis, but collaboration in the basic research proved a fruitful use of resources for the participants. Such a project might be considered on an international level.

As an aid to technical advance, ideas for educational programs were suggested at several panels. One type of program might involve the long-term assignment of a technical expert to a national or foreign university for specialized study. Jean Cantacuzène proposes that such a program can assist industry when requirements are made for the private sector to reconvert or assimilate new techniques. He cites as an example a program conducted through the Délégation Générale à la Recherche Scientifique et Technique for French industrial engineers to pursue studies related to enhanced oil recovery at the University of Texas at Austin.

The second category of information includes issues related to the structure of industry participation in cooperative agreements relative to the eight technical fields considered: Are existing conditions generally appropriate for the private sector to participate in cooperative arrangements? What arrangements are best suited to avoid conflict with national laws such as antitrust regulations, patent rights, and possible restrictions on the transfer of technical information? What role appears to be preferable for private sector involvement? For planning and recommending priorities? For conducting technical programs? For using and transferring results? Which of the following mechanisms would be most appropriate for accommodating a particular role?

- Exchange of specialists
- Sharing of equipment and/or materials
- Sharing of facilities
- Visits

- Research Task Sharing
- Exchange of technical data
- Conferences
- Cost Sharing

The chapters of this volume include considerable insight into the structure and type of participation appropriate to each technical field. Roland Schmitt's paper from the panel on materials offers particularly thoughtful comments on whether or not industry should be involved in activities conducted under international technical agreements and, if so, under what conditions. While his comments relate specifically to materials, they also reflect the consensus of opinion articulated at other panels. Dr. Schmitt summarizes his remarks with the following five main points that have application for a general consideration of industry participation.

1. If industrial benefits are to be gained from international technical agreements in the materials area, the industrial objectives should be kept in mind from the beginning of the negotiations and from the beginning of the conceptual formulation of the agreement. Thus, the first requirement is to understand and agree upon the real purposes of the agreement.
2. The delegations that negotiate such agreements should be staffed with people who understand and can deal with complex industrial issues that may be beyond the knowledge and experience of the experts in intergovernmental affairs.
3. It is important to understand and to consider the limitations and barriers to industrial sector participation, especially so as to avoid misleading the other parties in the negotiations who may have quite different constraints. Some of the factors that must be uppermost in the minds of U.S. negotiators are export control regulations, antitrust law, proprietary interests, patent rights, economic and business benefits, and cost and budgetary factors.
4. Where private sector participation is desirable, delegations should include people who have had some experience in managing such cooperative programs.
5. It would be highly desirable for the governments involved to see that their representatives bring with them a truly integrated set of governmental policies. In the United States, for example, the government itself should ensure the integration of the policies of the Commerce, State, and Defense Departments and should not require private industry to make this integration.

THE BOTTOM LINE

The motivation for addressing this subject arose from the recognition that productivity growth requires the optimum use of all resources. Those resources, both technical and financial, which are devoted to international technical cooperation are substantial and are increasing. How to link the strengths that such cooperation offers to the strengths of the private sector in a mutually beneficial leverage of resources follows no prescribed formula. As a first approach to the subject, this volume therefore does not contain a definitive map of the technical areas and mechanisms for linkage, where linkage is deemed useful and appropriate. What this volume does provide is a guide to an essentially unchartered territory of untapped resources which we may no longer have the luxury of leaving unexplored. Subsequent detailed considerations of this field will result in a more orderly approach to extracting the very considerable values contained in cooperative activities, which can then be expanded through the network of technical organizations, governmental and nongovernmental, throughout the world.

2

Opportunities for Improvement in Industrial Productivity

Jacques Desazars de Montgailhard

The initial topic proposed for this chapter was "opportunities for improvement in industrial productivity," with a suggestion that I attempt to illustrate the impact of scientific and technical progress on industrial development and the need to make the best possible use of technical resources, whether public or private, in order to ensure optimum efficiency. With this assignment, I must admit that I remained somewhat perplexed; my first reaction was that in a conference such as this, there was no need to enter into the justification of science, technology, and cooperation, since I am sure that we all consider the need for them as self-evident. But, on second thought, I came to the conclusion that it might be of interest to analyze the actual attitude of industrial concerns to scientific or technical progress and to cooperation.

I should first like to stress that I shall consider the subject only from the industrialist's point of view and not from a governmental one, since governments have political, national, and international interests which may lead them to decisions of a somewhat different nature from those which an industrialist might take in the same circumstances. Therefore, based on my experience in industry, I shall attempt to answer three questions: How is industrial productivity enhanced? What is the attitude of industrial concerns in cooperation? How can technical cooperation be expanded in different types of industrial activity?

HOW IS INDUSTRIAL PRODUCTIVITY ENHANCED?

To be competitive, an enterprise uses its own personal knowledge, which constitutes what I shall call its technical capital. This technical capital is the source of its strength in relation to other undertakings, and, in order to succeed in its struggle against competitors, the company must seek not only to protect this technical capital, but also

11

to enhance it. What then are the generators of progress? There are two: science and know-how. First of all, science – both fundamental and applied; this is a rational form of knowledge which, nowadays, is developed almost exclusively in public and private laboratories. The essential nature of science is an attempt to understand phenomena, and this search for knowledge for the sake of knowledge, even in its most abstract forms, has proved to be one of the most powerful factors in the development of humanity.

The second source of progress is know-how, which is empirical knowledge. Like science, know-how is the result of intelligence, patience, and the work of successive generations. But, since it does not seek to constitute a rational construction, know-how often opens up ways and means not yet attained by science.

Science and know-how often go hand in hand. Without going into examples with which all are familiar, it can be said that very many techniques or industries have an empirical basis and were subsequently developed by the contribution of science; others, on the contrary, owe their origin to a scientific discovery and have since progressed with the aid of experience.

It is often apparent that the most rapid developments occur when there is a balanced contribution of science and know-how. There is nonetheless a progressive tendency to give priority to science over know-how. But in my opinion, an attempt should be made by a suitable organization of companies to prevent the disappearance of know-how, whose "random" aspect often paves the way to progress which the overlogical scientific approach is unable to produce.

What, then, is the attitude of companies to these two sources of progress? The historical attitude has always been jealously to conceal know-how on the grounds that, once disclosed, it could be readily reproduced and be directly usable. Fundamental science, on the other hand, does not always offer immediate applications, and often justifies international cooperation for development. I shall revert to this point later.

To sum up, I should say that science is a form of logic and know-how is a form of art; once a scientific result is known, many people can find the way which leads to this result, whereas the ways of artistic invention are more mysterious. The trend toward cooperation increases when we move backward from know-how through applied science to fundamental science.

But how can an enterprise acquire these two generators of progress? We shall consider four ways:

- First, available information – publications, symposiums, and lapsed patents. This information is generally concerned with science and is normally free.
- Second, acquisition of knowledge – where this exists and where it is for sale. Examples are patents and licensing.
- Third, a company's internal research.
- Fourth, cooperation between a company and outside research organizations.

These last two ways, unlike the first two, provide access to new fields as yet unexplored.

These, then, are the generators of technical progress and the means open to companies to obtain access to them. As I have just said, cooperation can play an important part. But this raises another question.

WHAT IS THE ATTITUDE OF INDUSTRIAL CONCERNS TO COOPERATION?

The main idea guiding my analysis is that the aim of industrial enterprises should be to make a profit, or more generally, to achieve profitability. It follows that a company will or will not enter into cooperation depending on the profits it expects to draw from the cooperation. We must therefore seek the general factors in favor of or against cooperation.

It must be admitted that the adverse factors are numerous and weighty. I shall group them in two categories. First, the behavior of industrial concerns reflects human behavior which is fundamentally individualistic. This behavior implies the following: the tendency to keep to oneself what one has acquired; the desire to do everything oneself; and the search for supremacy. These behaviors are often ingrained, and while they are not conducive to cooperation, it must be admitted that they have a strong effort-stimulation effect and thus generate progress. Also, great care should be taken not to diminish the contributions of these behaviors.

Second, any cooperation encounters material difficulties which are further increased in the case of international cooperation by distance, language problems, cultural barriers, administrative and legislative difficulties, and so forth.

On the other hand, there are fortunately factors which are conducive to cooperation, such as: the impossibility of doing everything single-handedly, because of the absence of the requisite skills and the lack of human and financial resources; the possibility of undertaking more ambitious projects in cooperation; the reduction of costs and the division of risks; the savings in time; the benefits stemming from the multiplicity of skills and cultures; and also, I believe, a sentiment of human solidarity which often inspires teams of scientists and finds its concrete expression in the aid granted to the less developed countries. I am tempted to summarize this thought in two ideas: instinct and self-interest, on the one hand; and, paradoxically, interest and altruism, on the other. Instinct and self interest are a deterrent, and interest and altruism are an incentive, to cooperation.

HOW CAN TECHNICAL COOPERATION BE SET UP IN DIFFERENT TYPES OF INDUSTRIAL ACTIVITY?

I shall distinguish three areas – high-technology sectors, basic activities, and the area which might be called noncompetitive.

High-technology sectors

These activities pave the way for the future, and it is vital for every company to develop them. They have a number of characteristic features:

- They open up new fields and aim at satisfying new needs.
- Creativeness and research play a dominant role.
- The efforts required to achieve success are often long and costly.
- Technical supremacy is essential.
- It is a tight race to be the first to succeed.
- Often the initial market is very narrow and the price of the product very high, so that profitability is long in coming.
- The hope of profit is considerable in the event of success, but the risks of a setback are also very high.

As a consequence, there is considerable individualism and often severe competition. A typical example at the moment is photovoltaic energy. There is very little cooperation, except where it is powerfully motivated; such as when the cost of the project cannot be absorbed by any one company; when the complementarity of skills is essential; when the saving of time facilitates commercialization; or when the government provides incentives for cooperation.

One can argue whether such severe competition is profitable or sterile. I think it is difficult to make a general answer. It certainly leads to unnecessary effort, because the same efforts are made in several different places, but it leads to emulation which generates progress and almost always guarantees the quality of the solution that is ultimately successful.

High-technology sectors of activity often have the benefit of government assistance. I shall not discuss the judiciousness of this assistance; I shall merely stress that, inside each country, such support should leave ample scope for free competition between the companies concerned, especially in terms of research, even though it may be more directive when the time comes for selecting the best solution.

International cooperation can also be very fruitful in high-technology sectors. Here, however, governments generally are involved, due to either the size of the projects (nuclear energy and the space industry, for example) or the size of the financial requirements. The organization of this type of cooperation is therefore difficult. Moreover, it implies the ability to reconcile the implementation of an arrangement offered by governments with the constraints of politics, law, legislation, commercial interests, and the necessary freedom of companies in their management and technology.

Basic activities

This evokes, then, a consideration of basic activities. These include chemicals and metallurgy in general, or more specifically, the automotive industry. Our Pechiney Ugine Kuhlmann Group conducts most of its activities in these sectors, in which it is essential to make consistent advances in frontier research. If research is a guarantee of the future in the high-technology sectors, in basic activities it safeguards not only the present, but also, to a great extent, the immediate future.

Since basic activities are fundamentally concerned with processes, the essential problem is one of reducing costs, a prerequisite for remaining competitive. The progress made may be either continuous, based on a permanent effort across the board with respect to know-how, or discontinuous, involving radical changes in present processes or introducing new processes. In this second case, the gains are often fairly risky even though they may concern large quantities. Moreover, an old process based on considerable experience and having benefited from many improvements and developments which have made it reliable poses a substantial resistance-to-change factor. The risks are therefore high, and development for change requires time and heavy investments.

Very roughly, it might be thought that continuous progress would lead a company to specialization, while discontinous progress would prompt it to seek cooperation. In fact, experience shows that this is only true if the advantages arising from the association of skills, the size of the investments involved, and the division of risk and the prospect of better acceptance of the result by the partners outweigh the drawbacks of the need to participate in a program which is generally costly and likely to last over several years and the loss of independence implied by this cooperation. The choice between these two policies is ultimately a matter of circumstance.

But there is one form of cooperation related to processes which I should like to stress, and that is sales of technology — sales of licences, of know-how, and even of turn-key production facilities. Sales of technology are an extremely effective way of disseminating techniques. Independent of the fact that they are an excellent means for a company to make its own research profitable and enable the company to generate resources for the acquisition of other processes, they nurture an environment conducive to cross-fertilization. Some of the best illustrations of this are the mechanisms for technology transfer to less developed countries which can enable them to achieve a more advanced technical level than if they had relied solely upon their own resources.

Noncompetitive areas

Activities of the last category, which I have called noncompetitive for want of a better word, are also a theater of competition. This sector mainly concerns regulations of a technical, legal, social, or ecological nature and can affect the fundamental research in which industry engages.

With respect to the first point, the standardization of products and the differences in regulations and legislation in different countries often create artificial and protective barriers. Constructive cooperation involving industry and governments should enable us to break down these barriers. The same is true of social and ecological regulations, which should be more harmonized from country to country. In fact, these "noncompetitive activities" should lead to sound international competition, which is an undoubted factor of progress.

The same is true of fundamental research in which the desire for prestige often occasions a praiseworthy competition and profitable emulation. There has always been a very strong tradition of international cooperation in fundamental research, and industry plays a large part in this field. But in the difficult period which we are presently encountering, one feels that industry sometimes tends to withdraw into its shell — to keep to itself the results of its fundamental research, considering that these results may one day be useful in the struggle against competitors.

But I am convinced that fundamental research needs exchange and cooperation in order to flourish and that cooperation at the industry level is particularly essential to bring pure scientists back to a sense of the concrete. We all have everything to gain in maintaining a strong flow of international discussion between public and private research. Again, governments, with their aids, have an important role to play in promoting high-quality research.

In conclusion, I would like to stress a few points which seem to me necessary for sound cooperation. All the partners must feel that cooperation is the best means of achieving a given objective. The cooperation must be freely accepted. The benefits of the cooperation must be equitably shared among the partners, and the same should be true of any drawbacks. I shall close with a wish. There are already sufficient barriers to cooperation, and my wish would be that we not add artificially to these barriers.

3

Technical Cooperation and Industrial Growth: A Survey of the Economic Issues

Robert G. Hawkins

Technological cooperation may be positive or negative for global industrial productivity and both good and bad from the national point of view. As is true of most things in economic and political life, it all depends upon the conditions of the individual case and upon the perspective taken as the reference against which goodness or badness is judged. This chapter argues that technical cooperation across national boundaries is rising, but still remains at modest levels. It also maintains that, on balance, technological cooperation enhances industrial productivity.

The intent of this chapter is modest. It seeks to provide an economic explanation of the tendency toward industrial cooperation and its potential positive impact on world productivity growth. The need for technical cooperation, of course, stems from the division of the world into artificial and more or less arbitrary political units of nation-states. Without such borders, the issue of technical cooperation across national boundaries would be of no more interest than that of technology transfer from Hamburg to Mannheim. With sovereign nation-states, however, international technical cooperation and technology transfer are important because they involve the hurdling of the policy barriers at national frontiers.

The policy barriers become the focal point of the second intent of the chapter. This is to examine and to offer hypotheses to explain the role of national interests and policies in shaping and inhibiting technical cooperation across national frontiers. While more intensive cooperation may be very compelling from the economic perspective of the global economy, the economic and noneconomic objectives of individual nations may not be well served. Thus, selectivity in policy treatment at the national level, and the economics of individual industries, suggest which types of technological activities may be more fruitful for technical cooperation. This is a third objective of the chapter.

WHAT IS TECHNICAL COOPERATION?

Technical cooperation across national boundaries is not new. Examples for the twentieth century abound. And international technical cooperation can take many forms. Indeed, in the postwar period, it has grown by leaps and bounds, facilitated by dramatically declining international transportation and communication costs.

To delimit the scope of this chapter, it is necessary to exclude perhaps the most dramatic areas of technical cooperation among nations – albeit unconscious cooperation. Technology transfer by means of multinational enterprise activities of licensing, producing, and selling in various nations has been one of the most rapidly growing economic phenomena of the last thirty years.(1) This has done much to raise industrial productivity globally, although the benefits to individual nations are frequently questioned by local interest groups. A second major area of technical cooperation has been exchanges and meetings of academic and professional societies. These continue to serve as a basic medium through which new inventions, adaptations, applications, and theories are diffused internationally. The result is that ideas, problems, and technologies are addressed simultaneously in several national locations.

This chapter excludes these important areas of indirect technical cooperation. While multinational enterprises may produce cross-national technical cooperation within the enterprise, our attention is on technical cooperation between enterprises of different national bases. This scope is further limited by excluding agreements among companies for licensing of existing technology, acquisition of ownership of one firm by another, and production and marketing cooperation agreements. Thus, the rather narrow focus is on technical cooperation agreements among enterprises with different national bases, to conduct research and development, develop new products or processes, or otherwise collaborate in technology-intensive activities wherein the interaction is intended to increase the productivity of the outcome.

Taking this narrow definition, the frequency of technical cooperation agreements appears to be increasing, but is still relatively low. If we leave aside technical agreements in energy and the service sector, a survey of announcements of significant technical cooperation agreements for the year ending in May 1980 indicates that at least seventy-eight such cooperation agreements were made.(2) The criterion for inclusion was that the agreement covered either research and development or collaboration on the development of a new product or process.

While the number seems significant, it must also be viewed as modest. Some of the announced agreements will ultimately not come to fruition. Perhaps most important is that two-thirds of the agreements represented two industries – electronics and aircraft. Cooperation agreements in other manufacturing technology remain relatively rare.

Note that the data do not include the large number of technical and financial agreements involved in petroleum exploration and production.

These, of course, often involve companies of two or more nationalities, but normally involve the joint application of known technology to spread the risk and pool the technical manpower and capital. Nor do they include technology transfer to developing countries by national or international technical assistance agencies. While these may be viewed as instances of technical cooperation, they fall outside the scope of this chapter.

ECONOMIC BENEFITS OF TECHNICAL COOPERATION

The potential benefits from technical cooperation are quite simple: the development of new and better products and processes in a shorter time and with less resource utilization overall than would occur without cooperation. This, in turn, would result in higher productivity (output relative to input) in one or all of the cooperating countries.

The benefits may stem from several sources. One is that the pooling of technical talent and financial capital across national boundaries permits new projects to be undertaken which are beyond the means of any one of the participants. Such collaborations are undertaken to extend the frontier of technology or its application. Dramatic examples of such collaborations are in the aircraft industry, and the U.S.-Japan-Europe coal-gasification project in Colorado. New projects in some areas of industry, especially energy and extraction, have such heavy requirements of capital, of technical, engineering, and professional talent, and such high risk that no single enterprise or even nation can provide the full complement of resources. Thus technical collaboration across countries may bring products or resources into being which otherwise would have to wait. Almost always, such technical cooperation agreements have had the active involvement of one or more governmental agencies.

A second source of benefits from technical cooperation is the conservation of resources. Cooperation may avoid unnecessary duplication of research facilities and development labs; it may conserve scarce manpower; it may provide economies of scale in research and development efforts and facilitate the application of new innovations in several locations or product lines. Recent examples of technical cooperation, apparently with such intentions, are several. One is the proposed research and development collaboration among six European automakers; another is the Thomson-CSF/Xerox collaboration of videodisc technology.

Perhaps most important is that these collaborations to conserve and more efficiently allocate technical resources do not often seek to make major technological breakthroughs at the technical frontier, but rather seek to help the collaborators remain competitively viable in industries where technological leadership is held by other companies. Indeed, of the seventy-eight collaboration agreements in the sample, the circumstances reported suggest that about 70 percent were for the purpose of "catching up," or remaining competitive, with the perceived industry leaders.

If this is in fact the case, two hypotheses are suggested. The first is that the apparent intensification of international technical cooperation is the result of intensifying competition in world industries. There is considerable evidence that such has occurred over the last two decades.(3) In virtually every manufacturing industry (defined at the SIC 3- or 4-digit level), the world market share of the top three or eight enterprises is lower in 1978 than in 1960. The strongest competitive thrust has been from Japanese firms, whose world market share overall, and for many individual industries, has risen dramatically.

The second hypothesis suggested is that international cooperation agreements may serve to enhance or preserve competition. If, as the data imply, most collaboration agreements are associated with technological catch-up, they may be viewed as a logical response by companies to become or remain viable competitors with the technological leader(s) in the product or process. Both hypotheses have sound a priori bases, but require more conclusive evidence than is currently available.

Finally, technical cooperation agreements may be viewed as one mechanism in the harmonization of national industrial policies. However, this not yet a well-developed mechanism. National industrial policies, including the array of differential aids and incentives for private industry, the selective involvement of the public enterprises in particular activities, and differential free trade and protectionist policies across industry, are now characteristic of all industrial countries to some extent. A danger of industrial policies is that they can be ineffective and indeed countereffective if they do not take into account and become coordinated with the industrial policies of other countries. Thus far, this has proven to be a major difficulty, as industrial policies change in unpredictable ways. In any event, international technical cooperation, either by private enterprises or with government involvement, represents a potential means of achieving some level of coordination in industrial policy implementation – a largely unexploited one, however.

ECONOMIC COSTS OF TECHNICAL COOPERATION

While there are identifiable economic benefits from technical cooperation arrangements, there are circumstances in which such agreements are not an unmitigated good, and indeed are negative. It is in this analysis, however, that the perspective is important.

First, under what circumstances may industrial cooperation schemes be harmful for world productivity? One clear answer is: when the scheme solidifies or creates a true monopoly, providing the market power to raise prices, establishing barriers to the entry of competitors or stifling future technological change. While the conditions leading to this result are easily described, they are not so easily identified or predicted, in fact. Nor is there any presumption that this danger prevails only for technology cooperation agreements among private enterprises. There is abundant evidence that public sector companies and governments are not always timid in exercising monopoly power to extract gains from the rest of the world.

The second, and more complicated, perspective is that of the individual nation. For the nation, perceived harm may arise from technical cooperation arrangements even when the result from the world perspective is positive. This may occur for at least two legitimate reasons.

First, international technical cooperation agreements are likely to move domestically developed technology or technological potential to a foreign site or to joint ownership with foreigners. This implies some loss of national control over the technology. The loss of control may be harmful for certain domestic objectives. It may be viewed as an economic loss of a valuable intangible national asset with economic worth. In the more obvious cases, it may involve a loss of control over technology of strategic military importance (such as nuclear energy or state-of-the-art computers).

A less obvious type of loss for a nation results from the loss of positive "externalities" when the development of the technology occurs outside the country. These positive externalities, social or economic, may take several forms. One is to have highly educated, technically trained personnel located within the nation. They may stay after the project is completed, and are generally socially productive citizens. Another is that the results of the technical cooperation may be more immediately applied and have greater spillovers in the host economy than in other cooperating nations.

Governments, being more concerned with these "positive externalities," tend to view the location of the technology development activity with more importance than do private companies. If a technical cooperation agreement results in the location of most or all of the activity outside the nation, these external benefits are lost to the country even though the participating national enterprise may be very positively affected. Thus, the national policymaker is faced with a decision based upon the probability that similar activities would be developed locally if the technical cooperation agreement were not permitted. That these concerns are not trivial is attested to by the active involvement of governmental agencies in production-sharing agreements, especially in the aircraft industry.

In summary, the individual nation may perceive technology-sharing agreements as an inferior substitute to developing the technology locally, even when it is available at a lower economic cost from a foreign enterprise. The government must assess, first, whether the technology can be developed locally if technology cooperation agreements are disallowed and, second, whether the retention of control and the positive externalities to the nation from local development is worth the additional cost.

IMPEDIMENTS TO INTERNATIONAL TECHNICAL COOPERATION

The possibilities of national loss and the uncertainties over the gains from international technical cooperation combine to produce impediments to such arrangements. The most important such impediments arise from national policy interference.(4) But others exist as well, deriving from culture-bound management making biased decisions against cooperation with foreign-based enterprises, due to lack of understanding, mistrust of motives, and similar considerations.

Three rationalizations of governmental involvement can be identified. The first, which appears to be largely intractable, stems from noneconomic international military and political objectives of the nation. Prohibition or control of technical cooperation agreements now exist for every nation when technology important to national defense is involved, when the potential cooperator is perceived to be an adversary of the nation, or when an important national, social, or political issue is at stake (such as violation of human rights). Without commenting on the success of such controls, they are unlikely to disappear in the forseeable future and do constrain the growth of technological cooperation.

The second rationalization for involvement pertains largely to the U.S. context. It concerns the application of antitrust policies to preserve competition. There is little doubt that U.S. antitrust policies are perceived to limit international technical cooperation, based upon a series of earlier court decisions extending U.S. antitrust law to foreign activities which affect competition in the U.S. market. Despite great uncertainty and confusion about this issue among U.S. companies, one result is hardly debatable. American management is less anxious to seek out and enter technical cooperation agreements with potential competitors – in the U.S. or abroad – than would be the case if U.S. antitrust policy were differently oriented or had different precedents. To a lesser extent, this type of impediment operates in other countries.

A third motive for governmental involvement in international technological cooperation is to ensure that any international agreements further the nation's industrial policy. In some cases, the government may actively encourage international cooperation agreements, or a public enterprise may be one of the parties. This may occur when the nation's industrial policy calls for a production capability in a particular industrial sector and the existing viable technology is all in foreign hands (as with Honeywell-Bull in computers), or where any one nation's enterprises would find the project too large and risky (as with Air Industrie). In such cases, the international technology development or sharing agreement serves as a vehicle for the coordination of industrial policy across the cooperating countries. Such ventures appear to be economically beneficial from a global perspective by increasing the number of viable competitors in the industry, except when significant subsidization of a venture is required to keep it financially feasible.

In some instances, governments encourage technical cooperation as a matter of industrial policy to provide a failing industry (or company) with new product lines or technology to allow it to remain in business. (This may have been true of the original agreement between Machines Bull and General Electric.)* In such cases, it is unclear whether the arrangement has a net positive or negative effect on worldwide economic growth and efficiency. There is a definite potential for encouraging international linkages for a declining enterprise when other more fundamental changes would be more appropriate.

In other cases, governments actively interfere with or limit the scope of international cooperative agreements in technology as a means of pursuing these national industrial policies. This may occur when it is decided that the nation should have local capabilities or obtain leadership in the industry, should have the positive externalities associated with it, should avoid the economic costs of permitting an industry to decline, or for other reasons. In such instances, the government may forestall international cooperative agreements to secure a market for local enterprises through restrictive government procurement policies, import restrictions, or other measures. Further, preferential tax or credit treatment may also be made available for the designated enterprises.

The latter package of direct governmental initiatives most fully describes, of course, Japanese policy in such sectors as electronics and computers. But it also applies to other countries as well, in the same or other industries. The essential point is that a definite and positive industrial policy is not always (if ever) compatible with freedom for enterprises to enter any and all technology-sharing arrangements with foreign companies. The government must pursue a "second-best" policy of review and direction with respect to technical agreements to ensure consistency with the broader industrial policy.

BREAKING IMPEDIMENTS TO INTERNATIONAL TECHNOLOGICAL COOPERATION

The foregoing analysis suggests that international technological cooperation is generally beneficial from the perspective of the world economy. It permits large-scale projects to be undertaken which are beyond the capacity of any one nation or enterprise. It reduces duplication of development efforts and enhances competition by permitting competing firms to catch up and challenge technological leaders. Yet some forms of technology cooperation may be harmful when monopoly positions are created or solidified.

From the perspective of an individual nation, technology cooperation may be either positive or negative. Economic advantages of lower costs and more rapid technological progress are possible. But technological cooperation which adversely affects the national security of a nation or compromises other international economic objectives may be harmful. Further, cooperation agreements which apply the technology

---

*Honeywell later assumed GE's role.

in a foreign location may negate benefits of the positive externalities of learning by doing and of more rapid application of new technology for the individual nation. Overall, the conclusion is that such arrangements are likely to be beneficial in most cases, for both the individual nation and the world economy. Impediments to international technological cooperation are thus better removed than intensified.

Yet, how can governmental policy impediments to cooperative arrangements be removed? International cooperation arrangements may be segmented into three broad categories, for which the answers differ. Each category requires a different approach for addressing potential impediments to technological cooperation.

The first category involves technology which affects a nation's perceived vital national interest, that is, its military and security position. Little can be done to remove existing impediments in this area. If governments expect benefits from international agreements involving potential weaponry or defense capabilities, they will be negotiated on a government-to-government basis. There is little likelihood that this practice can be significantly altered by international agreement.

A second type of potential cooperation arrangement involves very large, technologically intensive and expensive projects. Examples include supersonic air transportation, deep seabed mining, and the commercial exploitation of outer space. Viable ventures in these fields are normally beyond the capacity of one enterprise and often beyond that of a single nation. It is also frequently the case that property rights beyond their defined international boundaries are involved. It is further the case that one or more governmental agencies are involved for each of the collaborating nations.

One result is that governmental involvement, directly or indirectly, is required in any international technological collaboration of such massive scale. Thus, negotiating international agreements must involve the appropriate governmental agencies, preferably at early stages. When the interests of many nations are involved, a multilateral international agreement clearly has advantages in setting the parameters within which technology-sharing arrangements between governmental and private enterprises can be negotiated.

The third and most important set of impediments includes those which limit the freedom of technology cooperation among private enterprises on "nonfrontier" projects. There are potential economic benefits from enhancing effective competition and the avoidance of unnecessary duplication of research and development efforts among interested companies. Governmental interference in international collaborative agreements in these cases stems from several possible economic and political motives. These include the objectives of creating domestic jobs, avoiding balance-of-payments drains, and views about the location of scientific and production facilities.

Removal of impediments in this area may involve several possible actions. One, of course, is for the potential collaborators to structure the cooperation agreement in a way that the economic interests of the national governments are considered. An innovation is the international production-sharing agreement, in which production of components, assembly, and engineering and development are distributed among various national locations to reduce the adverse impact on national economic objectives of the private international collaboration.

Yet governments also have potential benefits in removing impediments to international technology-cooperation agreements, since the flexibility to "go around" national policies in private agreements is limited. But a major problem in removing such impediments rests with the nature of international economic policymaking.

A technology-cooperation arrangement involves international capital movements, exports and imports, and a flow of human resources. Yet the international economic system deals with balance of payments and international finance on one policy plane and through one set of institutions. International trade policy, on the other hand, is dealt with on another plane and in a different set of institutions. To accommodate technology-cooperation arrangements, these planes and institutions must have a much higher level of intersection than currently exists. Only then will the full economic potential for international technology cooperation be realized.

NOTES

(1) See, for example, Edwin Mansfield, "Technology and Technological Change," in Economic Analysis and the Multinational Enterprise, ed. J. Dunning (London: Allen & Unwin, 1974); and Arthur Lake, "Technology Creation and Technology Transfer by Multinational Firms," in The Economic Effects of Multinational Corporations, ed. Robert G. Hawkins (Greenwich, Conn.: JAI Press, 1978).

(2) The survey included individual cooperation agreements publicly announced and reported in Financial Times, The Economist, and World Business Weekly. The survey is obviously incomplete, but provides a rough indication of the frequency and type of agreement pursued.

(3) Specific evidence is contained in Raymond Vernon, "Competition Policy Toward Multinational Corporations," American Economic Review, May 1974; and John Dunning and Robert Pearce, Profitability and Performance of the World's Largest Industrial Companies (London: Financial Times, 1975). For a review of the evidence, see Robert G. Hawkins, "Are Multinational Corporations Depriving the United States of Its Economic Diversity and Independence," in The Case for the Multinational Corporation, ed. C. Madden (New York: Praeger, 1977).

(4) Thomas Gladwin and Robert G. Hawkins, "Conflicts to the International Transfer of Technology: A U.S. Home Country View," in Technology Transfer Control Systems, ed. T. Sagafi and R. Moxon (New York: Arts and Science Press, 1980).

4

Scope and Function of International Technical Agreements*

Klaus-Heinrich Standke

The network of scientific and technological relations between states, between public agencies, and between firms still appears to be relatively little known despite many studies in recent years. This network is the sum of multivarious links which are often confidential, such as cooperation agreements between private firms, and which, moreover, are in constant evolution.

Behind the work carried on in technical cooperation, we find governmental bodies (national and international), industrial or commercial enterprises, universities and non-profit-making organizations and associations. Hence, any analysis of the scope and functions of international technical agreements must be preceded by a study of the organizations which stimulate and set in action cooperative work. Thus a better understanding of the functioning of the international organizational machinery will make an analysis of international technical agreements more meaningful. This chapter is therefore designed as a reference for these organizations and agreements.

The bilateral exchange of people or of publications has been for many centuries the only form of international scientific cooperation. Such cooperation on a multilateral scale is a development which started only in the nineteenth century on themes which are called today "global problems," such as meteorology, astronomy, and geophysics. "International years" for the study of selected scientific problems have been organized, calling for the international cooperation of scientists from different disciplines. For example, the first International Polar Year was organized in 1882 with the active participation of eleven national expeditions and observers from thirty-five other countries.

Another form of international scientific and technical cooperation was the organization of international congresses. The first World

---

*The views expressed in this chapter are those of the author and do not necessarily reflect those of UNESCO.

26

Economists Congress was held in 1847, followed by international congresses on agriculture (1848), sanitary issues (1851), meteorological observations on the sea (1853), statistics (1853), ophthalomology (1857), chemistry (1860), geology (1862), and so forth. International conferences have been and continue to be the most visible form of international cooperation.

The desire to institutionalize some of these ad hoc meetings in some sort of permanent platform gave birth to the creation of international organizations. The Bureau International des Poids et Mésures in the Parc de Saint-Cloud (France) is the oldest of these nongovernmental organizations. In 1900, on the initiative of the Academy of Science in Göttingen, the International Association of Academies was created which ultimately led to the establishment of the International Council of Scientific Unions (ICSU), grouping practically all scientific disciplines.(1) The creation of the League of Nations after the First World War, though on a modest scale, allowed for an intergovernmental cooperation on selected scientific and technical issues.

The present status of international machinery is mainly the result of international initiatives after the Second World War. The creation of the United Nations System with its numerous specialized agencies has largely shaped the present global intergovernmental pattern. In addition to the United Nations agencies, a variety of regional organizations have also been created.

In parallel to these institutionalized intergovernmental initiatives, there has been large-scale development of various other kinds of international scientific and technological relations over the last thirty-five years. With the involvement of government authorities, scientists, and industry, a host of new international cooperative activities – on both a bilateral and a multilateral scale – have been initiated. What these arrangements have in common is that the partners bring together resources on a bilateral or multilateral, permanent or temporary basis, with a view to achieving an advance in scientific knowledge or technological progress.

The boom in cooperative research has taken diverse forms. For instance, the sphere of competence of international organizations whose activities were originally of an economic nature has been extended into the scientific and technological field. This has been happening, in particular, to the Council for Mutual Economic Assistance (CMEA), to the European Economic Community (EEC), and to the Organization for Economic Cooperation and Development (OECD).

Other international organizations which have been active in the scientific sector from the outset have been giving greater importance to cooperation in research and development than in the past. This applies in particular to UNESCO (International Oceanographic Commission (IOC), Man and Biosphere Programme (MAB), International Geological Correlation Programme (IGCP) and International Hydrological Programme (IHP)) and various specialized agencies of the United Nations System, such as the World Health Organization (WHO), the World Meteorological Organization (WMO), the International Telecom-

munications Union (ITU), the International Atomic Energy Agency
(IAEA), and the Food and Agriculture Organization of the United
Nations (FAO). More recent organizations such as the United Nations
Industrial Development Organization (UNIDO), the United Nations
Environment Program (UNEP), the United Nations University (UNU), or
the United Nations Center for Human Settlements (HABITAT) have
given an important place to cooperation in research since they were
established.

The many activities undertaken at the multilateral level have been
steadily complemented by an increasingly dense network of bilateral
agreements between governments, between industrial firms, and between scientific bodies.

This spectrum of activities has been accompanied by extensive
changes in methods of organizing and managing concerted research.
Thus, the spontaneous collaboration still practiced by the scientific
community through the organization of congresses, missions, and meetings or the dissemination of publications has been reinforced by the
forging of direct links between major public agencies, between laboratories, and between research teams. Similarly, alongside the exchange
of patents and know-how between industrial firms, there is now close
cooperation between design offices in the development of new products
or new manufacturing processes.(2)

TYPES OF INTERNATIONAL ORGANIZATIONS

It is not an easy task to develop some meaningful classification of the
multitude of international organizations and, thus, of the multitude of
international technical agreements handled by them. The Economic
Commission for Europe (ECE) of the United Nations has made an effort
to systematize these arrangements and they are discussed briefly,
below.(3),(4)

Classification according to the legal status and membership of agency responsible for organizing the cooperation

To draw a complete catalogue of international organizations whose
statutes refer in one way or another to science and technology is not an
easy matter, and the absence of a concise worldwide directory demon-
strates this difficulty vividly. In some organizations, the science and
technology activities are dominant, while in others, those activities are
derived from assignments of a more general nature with an economic or
social slant, or are focused on such concerns as agriculture, health, or
culture.(5)

It is, however, useful from the jurisdictional point of view to make a
distinction between: (a) intergovernmental science and technology organizations (IGO), which are established by an agreement between
governments and whose membership consists of states or governments;
and (b) nongovernmental science and technology organizations (NGO),

Table 4.1. Distribution of Total Budget Allocations
for Science and Technology by Organization

1978-1979, ($000)*

Organizations and Agencies	Regular Budget	Extra-Budgetary
United Nations	$3,648.4	$44,284.9
ESCAP	472.0	618.0
ECA	362.0	475.0
ECWA	1,554.0	863.0
UNCTAD	1,326.0	150.0
UNIDO	512.0	2,876.0
UNEP	5,438.4	11,933.4
HABITAT	324.7	26,241.0
UNDP		37,895.1
UNITAR	38.0	1,963.0
UNU	5,425.0	326.0
ILO	13,254.0	54,866.0
FAO	26,507.0	17,147.7
UNESCO	15,288.1	24,818.0
ITU	0.0	5,215.7
WMO	12,167.0	8,100.0
IMCO	0.0	35,034.0
WIPO	4,195.0	2,939.0
IAEA	36,959.5	6,886.0
Subtotal:	$127,471.1	$282,631.8
WHO	208,288.0	194,061.0
Total:	$335,759.1	$476,692.8

*Data taken from United Nations, General Assembly, A/CONF.81/PC/CRP.13/Add.1.

which are generally formed as a result of nongovernmental initiatives that often proceed from the international scientific research community (their membership does not include states or governments).

The classification according to legal status of the agency responsible for organizing the cooperation is, however, of little operational utility if it is not considered in conjunction with the type of international instrument or contractual modality governing the cooperation.

Classification according to the type of international instrument or contractual mechanism governing the cooperation

Depending on the functional form of multilateral cooperation, the type of international instrument or contractual mechanism governing the technical cooperation may vary widely. For example, special inter-governmental agreements are usually used when creating new inter-governmental organizations enjoying an independent and international status. Programs and budgets specifically devoted to cooperative science and technology ventures may be adopted by the governing organ(s) of existing intergovernmental organizations. Contractual agreements may be developed between the executing international organization in charge of a cooperative science and technology project and the partners in the cooperating countries concerned. Legal incor-poration of a nongovernmental international organization may be used in a host country whose national legislation includes provisions for such formalities.

Classification according to the geopolitical area in which the coopera-tion takes place

A rough distinction can made between the following:

- Cooperation involving both East and West, which often – but not always – coincides with worldwide programs conducted under the aegis of UNESCO, WHO, WMO, IAEA, FAO, UNIDO, and others
- Cooperation involving the East, as exemplified by programs under-taken under the aegis of the Council for Mutual Economic As-sistance (CMEA)
- Cooperation involving the West, as exemplified by the programs sponsored by the Commission of the European Communities (CEC), the Organization for Economic Cooperation and Development (OECD), the International Energy Agency (IEA), the North Atlantic Treaty Organization (NATO), the Council of Europe, the European Organization for Nuclear Research (CERN), the European Molecular Biology Laboratory (EMBL) and the European Space Agency (ESA).
- Cooperation involving regions of developing countries – the Organi-zation for African Unity (OAU), the League of Arab States, the Association of South East Asian Nations (ASEAN), the Association for Science Cooperation in Asia (ASCA), the Organization of American States (OAS), the Andean Pact, and so forth
- Cooperation involving one or several developed countries and a selected group of developing countries; for example, the Common-wealth Secretariat or the Agence de Cooperation Culturelle et Technique, or the Organization of American States (OAS).

The above classification covers exclusively intergovernmental or-ganizations. However, for the purpose of international scientific and technical cooperation, the nongovernmental organizations listed below are of equal importance:

Worldwide

International Council of Scientific Unions (ICSU)
World Federation of Engineering Organizations (WFEO)
World Federation of Scientific Workers (WFSW)
World Association of Industrial Technological Research Organizations (WAITRO)
International Federation of Institutes for Advanced Studies (IFIAS)
International Institute for Environment and Development (IIED)
Confederation of International Science and Technology Organizations for Development (CISTOD)
African Association for the Advancement of Science (AAAS)

Regional

European Industrial Research Management Association (EIRMA)
Industrial Research Institute (IRI)
International Institute for Applied Systems Analysis (IIASA)

Classification according to the degree of vertical integration of the cooperative projects

Here one can distinguish conveniently a number of phases in science and technology projects – mission-oriented activities which must be carried out in a limited time-span:

- The feasibility study phase (identification, selection and planning)
- The implementation phase (actual execution of R&D or scientific service activities)
- The phase of sharing and dissemination of results
- The phase of application into productive activities
- The phase of effectiveness evaluation.

Multilateral cooperative projects in science and technology can accordingly be grouped into categories according to the degree of their "vertical integration" of one of more of the five phases descsribed above.

In some cases of the past, selection of projects had been made independently by the partners who only later decided to join forces by coordinating the execution, to make joint use of results. The European Atomic Energy Community (EURATOM) and various interagency projects provide examples of this.

Sometimes partners decided to limit the cooperation to the selection of projects, and shared dissemination and/or application and/or evaluation of results. The purpose here is to give partners the maximum freedom of action by leaving them full responsibility for the execution of the science and technology activities. Such a pattern is used by some transnational enterprises and by certain intergovernmental organizations like UNESCO and FAO. Complete "vertical integration" covering all the five mentioned phases is very seldom.

A more pragmatic – and in any event, shorter – typology which may perhaps be more relevant for our needs here can be found in a working group publication of the European Industrial Research Management Association (EIRMA) analyzing various criteria of cooperative international research.(6)

Classification according to the intended missions

International organizations in science and technology fall into two categories – intergovernmental (about fifty organizations) and nongovernmental (about 250 organizations). From a functional point of view, a distinction should be made between organizations favoring exchanges of information, such as the International Council of Scientific Unions (ICSU); normative organizations, such as the International Organization for Standardization (ISO); and operational organizations, such as the European Organization for Nuclear Research (CERN), the European Space Agency (ESA) and the International Atomic Energy Agency (IAEA). Those which have been working with success have two types of function: on the one hand, the establishment and operation of research facilities combined with internal research and development efforts; on the other hand, the functioning as a contracting agency for cooperative industrial R&D.

In most organizations, the mission has been a mixture of these two tasks; the organization has primarily involved the function as contracting agency for industries in different countries for research and development for an ambitious project, combined with in-house R&D efforts. This mixture is presumably the best solution, since the additional in-house efforts and operations of facilities by the agency's own staff may guarantee the necessary continuity, scientific expertise of the staff, sensitivity to the technical problems, and training ground over many years.

SCOPE AND FORMS OF COOPERATION AGREEMENTS(7)

Cooperation is the process of pooling or coordinating the efforts of the cooperating parties to attain objectives which are set by agreement and which, depending on the nature of the work under consideration, may be scientific (improvement of knowledge) or technical (utilization of knowledge for practical purposes). The three main stages of this process can be distinguished as follows: (1) the preparation and selection of projects (the decision-making stage); (2) the execution of the work (the implementation stage); and (3) the dissemination and practical application of the results (the utilization stage).

Joint Preparation of Research Projects

Any new research project, no matter in what form it is to be carried out, requires preparatory work in order to establish the validity of the project and the conditions necessary for its execution. The problems posed by the preparation of researach projects are often of such a magnitude that extremely elaborate concertation machinery is required. A project is adopted when the objectives – scientific, technical, or economic – of the proposed operation have been defined and when the methods of attaining them have been specified. The adoption of any cooperative project of a certain degree of complexity is the result of a process involving several stages through which the partners define the content and scope of their respective commitments.

The central feature of this process is the definitive agreement which establishes the respective obligations of the parties. The agreement generally takes the form of one or more contracts which determine:

- The object and nature of the research (namely, fundamental, applied, technical, or semi-industrial)
- The period of validity of the contract and the duration of the corresponding clauses (in particular, the clauses regarding secrecy)
- The contributions of the partners and the corresponding remuneration
- The liability of the parties
- Definition of information and arrangements for patents which may have been obtained from third parties
- Methods of settling disputes

Cooperative research contracts involve pooling resources, capacities, and aptitudes in order to obtain results which are often hard to define in advance and which are by their very nature uncertain; this means that the partners need a high degree of flexibility in determining their respective rights and obligations. Any operation which is at all complex therefore includes a number of clauses relating to methods of evaluation and regular review of work; mechanisms for terminating or annulling the contract in the event of special difficulties or expenditure in excess of the estimates; obligations for secrecy in the event of annulment; and conditions for utilization of the results already obtained.

Cooperation in the preparation and implementation of these agreements entails close collaboration, possibly justifying the adoption of extremely varied procedures, between the parties concerned.

Joint Execution of Research Projects

Joint execution of R&D projects or programs is rightly regarded as the keystone of cooperation in research. It may take many different forms, the most important of which are establishment of joint permanent or temporary bodies; coordination of tasks among independent institutions or bodies; and systematic exchange of information on work in progress on the basis of a concerted plan.

Establishment of joint bodies

The formation of a joint body entails bringing together, for a specific research program or topic, staff from different organizations or countries. Whether or not it constitutes a legal entity, the joint body established in this way may be an institute, a laboratory, or, more simply, a permanent or temporary research team. Joint bodies are usually set upon the basis of a contract or statutory act which defines the laboratory tasks, the powers of the governing council, the method of financing, and the conditions under which the body can be dissolved. The following are among the forms of cooperation most frequently employed:

- The establishment of international research teams or laboratories
- Regular rotation of staff between the laboratories of the cooperating organizations
- Assignment of staff or research teams to foreign laboratories for execution of a specific project (or, conversely, invitation of foreign staff or research teams to national laboratories).

Exchange of information, according to a concerted plan, on work in progress

Whatever the mechanism chosen, cooperation through the division of tasks is an operation of which neither the direct nor the indirect costs should be underestimated, especially when the tasks have to be shared as fairly as possible among scores or even hundreds of participants.

A recent estimate of the cost of cooperative research at the national level was 10 to 20 percent of total commitments. Although similar estimates at the international level are very rare and not entirely objective, it would seem that this proportion is substantially exceeded in large international projects or programs.

It is for this reason that when costs seem likely to become too high, the partners sometimes decide to specialize in the areas of research in which they are the best qualified, entering into an agreement for the exchange of technical information (sometimes accompanied by the establishment of a patents pool) and ensuring that the cooperating body has full access to its partners' laboratories.

The definitive decision in the matter depends on the results desired through cooperation. The results in turn depend on a number of factors including dissemination of scientific and technical information, assignment of intellectual property rights, and forms of joint utilization.

Dissemination of Scientific and Technical Information

Dissemination of information takes place when reports on work completed or in progress are presented, or when links are forged between research workers and users on the occasion, for example, of in-service training periods, working meetings, or laboratory visits. In the case of technological research, dissemination may also take the form of sending specimens of industrial machinery or equipment for testing.

Cooperation in this field of dissemination may be the result of establishing a pool of information (with or without legal personality) in which the partners are equally represented, or, more simply, it may result from the dissemination of research findings according to agreed plans and methods.

Assignment of intellectual property rights and transfer of know-how

Assignment of intellectual property rights (patents, utility certificates, trademarks) takes place when the parties agree to grant to third parties a license which may, and often does, include the transfer of the corresponding know-how. The term "license" means the authorization given by the holder of an exclusive right (the licensor) to another person (the licensee) to perform certain acts covered by that exclusive right. The granting of a license is very often accompanied by contracts covering technical assistance and the supply of certain equipment.

The granting of a license entails reaching agreement on the following matters and incorporating provisions thereon in the corresponding legal instruments:

- The nature of the benefits which are to be transferred
- Any restrictions on the transfer
- The price to be paid for the transfer
- The procedure for payment
- The distribution of the profits (or other advantages) resulting from the operation
- The procedure for the settlement of disputes

Various procedures exist to reconcile the interests of enterprises or laboratories which have collaborated in developing techniques for which they intend to grant a license.

One possible solution, when the result has led to a patentable invention, is to take out a joint patent in the name of all enterprises or all design offices associated in the cooperation contract, in which case the assignment of rights would automatically imply the agreement of all the licensors. Another solution, which is common in all cases where no patent is taken out or when the patent is taken out in the name of one of the parties, consists in stipulating that assignment of rights is subject to the agreement of all the parties concerned and to their receiving the benefit according to a scheme of apportionment specified in the contract. In cases where the invention or technical innovation is essentially the result of research carried out by one of the laboratories or design offices concerned, it may be stipulated that a design office is free to dispose of such part of the technical information as it alone has developed; the principle of joint decision and the possible sharing of earnings would then only apply to improvements resulting from technological cooperation.

Joint utilization and capitalization of research results

The results of research are put to use when inventions or improvements to existing products and processes are exploited commercially. Commercial exploitation of the results of research passes through two stages, on the basis of which the corresponding forms of cooperation may be defined. The first stage is that of manufacture; the second, that of joint marketing of the products. When the invention which gives rise to cooperation has not reached the stage of effective exploitation, production and marketing are sometimes preceded by a development and testing stage which enables the partners to capitalize the results of the research on a joint basis.

THE LEVELS OF COOPERATION

The level of cooperation depends on the number of joint activities carried on. These activities, which were described above, may be concerned with the preparation and adoption of projects, the execution of projects, and the utilization of the results.

Depending on whether the cooperation is to embrace two or three kinds of activities, several different levels of cooperation may be distinguished corresponding to four situations:

1. Selection of projects + execution of research + utilization of results
2. Selection of projects + execution of research
3. Execution of research + utilization of results
4. Selection of projects + utilization of results

The first of the possibilities listed above represents the highest level of cooperation; it entails complete integration of the functions connected with the research-and-development cycle and its industrial sequels. Possibilities 2, 3, and 4 correspond to an intermediate situation which may be described as representing a part or limited integration of projects.

The Higher Levels of Cooperation in Research

Cooperation at this level may apply either to industrial operations in which a number of enterprises join together, usually for a long period, to develop a given "project," or to major international programs which are drawn up and implemented by governments and which lead to the establishment of an interstate body.

Cooperation between enterprises

This mode of cooperation brings several industrial entities into close association to produce a new piece of equipment or a new product. It is characterized by joint performance of all functions connected with the planning and execution of the activities which are the object of the cooperation. It may briefly be described as follows:

The main decisions, particularly those relating to the specification of equipment and the planning of research and testing, are taken jointly. A cooperating body, on which the participants are equally represented, makes the decisions and supervises their execution (preparation and adoption of projects).

Studies, tests, and updating activities are shared between the partners or carried out in joint bodies set up by them (execution of work).

The manufacture and marketing of the products (including after-sales service) are carried on according to the same principles: by apportionment of tasks among the partners or by the setting up of joint subsidiaries (utilization of results).

Cooperation between states

Integration is achieved through the procedures used by states at the different stages of the process of planning and executing research projects. These procedures present the following characteristics:

Research programs are drawn up by a scientific committee composed of eminent persons appointed or approved by the participating states.

The execution of the work is entrusted to a joint body set up by the cooperating parties; it may carry out the research in its own laboratories or delegate it to outside laboratories as the situation requires.

The agreements concluded between the participating states or public bodies lay down the terms on which information may be disseminated outside and on which each party may use the results of the research for its own needs. A joint operating company is often set up to undertake the industrial and commercial utilization of the results.

Cooperation Limited to the Selection of Projects and the Execution of Research

Cooperation between enterprises

The characteristic of this form of cooperation is that each partner is free to use the results as desired. There are three sets of circumstances in which this course may be chosen:

- The results can be put to use at a low cost which each partner can bear at little or no risk.
- Utilization of the results is an expensive operation and each partner prefers to take the risks separately.
- Utilization of the results is of interest to one partner only.

In general, the forms of cooperation just described seem essentially to meet the needs of enterprises wishing to retain maximum liberty when drawing up an agreement, on the assumption that at a later stage it is always possible to supplement the original contract by more stringent stipulations with regard to utilization of the results.

Cooperation within the framework of intergovernmental organizations

The extreme flexibility of this type of cooperation makes it an ideal instrument for collaboration within the framework or under the auspices of large intergovernmental organizations. It is well adapted to research of long-term and diffuse profitability, for which it is difficult to forecast in advance the conditions governing the transition to industrial application.

Cooperation Limited to the Execution of Research and Utilization of the Results

This situation is encountered where the partners decide to coordinate research programs which essentially have already been planned, and to make joint use of the results. The coordination of work generally results in an apportionment of tasks which may or may not entail modification of the programs as originally planned. The forms of cooperation differ, according to whether projects are undertaken on the enterprises' initiative or whether intergovernmental programs are involved.

Cooperation between enterprises

According to the already mentioned EIRMA working group report, No. 9,(8) this kind of collaboration is of interest mainly to enterprises whose activities complement one another. This would be the case, for example, where one firm supplied equipment to another and the two engaged in joint research at the design-office level, in some cases setting up a patents pool.

If it is to be effective, this kind of cooperation, which sometimes leads to technology subcontracting, presupposes a number of conditions. The ideal partner would be a large diversified company, since, even if work must be concentrated on a specific sector, it could receive valuable help from other sectors as well. The number of partners should be limited. They should have sufficient financial resources to enable them to assume the risks inherent in this kind of cooperation. Although similar in size and resources, the partners should have their strong points in different sectors or subsectors.

Cooperation within the framework of intergovernmental organizations

This form of cooperation, while less often found at the intergovernmental level, is used by international organizations wishing to enlist enterprises or research establishments of their member countries in the execution and utilization of research work which is decided upon and launched primarily at the national level.

Joint execution of projects. In return for substantial participation in financing the work (usually amounting to some 35 to 40 percent of the total cost), the cooperating organization – in this case, the Commission of the European Communities – acquires the right to select, under these contracts, both its own staff and approved staff of other European Community enterprises. This assignment of personnel to the contractor provides training for research workers and technicians and at the same time enables the Community to acquire the essential know-how which is not to be found in reports and patents. Supervision of execution is the responsibility of a management committee composed of representatives of the commission and the associated organization.

Dissemination and utilization of results. In return for the financing provided and for participation in the execution of the research work, the results obtained and experience gained may be circulated to all the enterprises and research establishments of the Community on the terms specified in the contract of association. These contracts usually include clauses providing for the grant of free licenses for the benefit of the commission and provisions against the freezing of information for the contractor's sole gain. With regard to patented inventions, it is usually provided that the contractor shall, if desired, hold the patent in all countries and that the commission shall have a free license for the purposes of its own program. It may grant sublicenses only to enterprises of the Community, save as otherwise provided by cooperation agreements with third countries. Furthermore, the commission may oppose the grant of licenses by the contractor if this is contrary to the interests of the Community. Detailed provisions define the conditions under which transfer of knowledge and know-how may take place.

It is for this reason that in practice those responsible for financing research often prefer procedures which require less funding by the federating body.

Cooperation Limited to the Selection of Projects and Integration of Results

The purpose of this last form of cooperation is to give the partners the maximum freedom of action by leaving them full responsibility for the execution of research, whether or not the cost is shared.

This form of cooperation is very flexible inasmuch as it leaves the participants as independent as possible. It is applied where the partners are so numerous that it is difficult to share the work. Its use is spreading rapidly in the majority of advanced countries, for it meets the needs of enterprises and industrial groups that are anxious to promote cooperation between research establishments scattered, in many cases, over several countries, and the needs of states that wish to encourage flexible and efficient forms of scientific and technical collaboration. One of the advantages of this kind of cooperation lies in the financing procedures which make each contracting body responsible for funding the work assumed by it, under the terms of the contract or agreement signed.

Cooperation between research establishments belonging to the same enterprise or industrial group

The magnitude of the research potential possessed by large enterprises – in particular, by multinational industrial groups – and the number and diversity of the research establishments under their control call for methods of planning and management which safeguard the initiative and independence of each center to the fullest possible extent: close collaboration at the decision-making level; broad autonomy at the level of execution; and pooling of results.

Cooperation between governments or intergovernmental organizations wishing to coordinate their activities

This type of cooperation is most often practiced, and perhaps most fruitful, at the multilateral level. It recurs, in various forms, in the many "concerted" or "coordinated" research programs undertaken under the auspices of CMEA, the European Communities (ECSC and ÈEC), OECD, and the organizations attached to them.

The "European concerted actions" prepared and undertaken since 1959 by Scientific Research and Technical Policy Group (PREST) and more recently by European Cooperation on Scientific and Technical Research (COST) appear to provide the best illustrations of a form of cooperation which is constantly being improved. In the absence of detailed information on the nature and scope of the agreements drawn up and implemented under the COST program and under the auspices of the organizations referred to above, this discussion will merely give, by way of illustration, some general information on the structure of "European concerted action in the field of metallurgy" – the only cooperative research project about which some fairly specific information has been published. This action, which was decided on by governmental agreement on November 23, 1971, and in which, in addition to ECSC, nine Western European countries are participating, has as its purpose the technological study and development of superalloys and alloys of titanium for gas turbines.

Adoption of projects. The selection of projects is always made in Community bodies. A management committee composed of representatives of the signatory states makes recommendations on the research proposals submitted to it, and states the grounds for its recommendations. In the committee, each representative has one vote. Procedural decisions are taken by a simple majority. All other decisions are taken unanimously.

Execution of the work. The cooperating parties are fully responsible for the execution of the work, and finance it in accordance with agreed scales of contributions. The signatory states conclude the necessary contracts with management of their performance.

Dissemination and utilization of results. The signatories insert into the contracts a clause requiring industrial enterprises or research establishments to submit periodic progress reports and a final report. The progress reports have a confidential distribution which, in so far as they contain detailed technical information, are restricted to the signatories and the management committee. The final report, which is intended to provide an account of the results achieved, is given a much wider distribution including − in the terms of the agreement − at least the interested industrial enterprises and research establishments to which the participants in the action belong.

Furthermore, the holder of the intellectual property rights resulting from the research is required to grant licenses for utilization on fair and reasonable terms where the needs of the market in the territory of the signatory applying for the license are not satisfied, or where the aim is to meet the domestic needs of the signatory applying for the license in matters of safety and public health.

Like any research program on a certain scale, the concerted "action" and "projects" just described fit into a dynamic process in which the decision makers are constantly having to review the choices they have made in the light of fresh experience. A project may thus result in more advanced forms of cooperation involving, for example, the establishment of a new legal entity or, alternatively, development towards more limited collaboration, such as the simple exchange of technical information. It may also vanish, either for good or temporarily, or give place to a new project.

A particular effort to discover methods of improving the efficiency of organization and management would be appropriate. Slowness in the preparation of certain programs, the complexity of decision-making procedures, and the underestimation of cost or time required for completion of the work are general problems involved in the management of all high-risk activities at all levels, national or international, individual or collective, in an integrated framework or with simple association of the partners.

MOTIVATION FOR AND AIM OF INTERNATIONAL, SCIENTIFIC AND TECHNICAL COOPERATION

Why, one might ask, after reviewing all these complicated bureaucratic procedures, is there a great and even an increasing interest in international cooperation?

From an industrial point of view, the five main considerations below can be stated.(9) Governments would most probably indicate the same motivations, although the choice of one or several cooperation partners tends to be primarily influenced by political concerns and less by technical or commercial considerations.(10)

Exchange of Information

The initial motive for cooperation across national borders has often been the exchange of ideas and information. The information obtainable through such contacts is of paramount importance as a stimulant for new ideas and thus for scientific and technological progress.

If exchange of information is the limit of the cooperative action, then only a loose association is needed, with little or no arguments against it and relatively few complicating problems.

Sharing of Resources

One of the main reasons for entering into international technical cooperation is the need to share various kinds of resources in finance, know-how, personnel, equipment, and others. Sharing of resources will primarily be desirable for economic reasons, although political reasons may in some cases be important.

R&D and capital expenditures for very complicated systems or large technology programs can be so high that the task cannot be solved within one country. These projects, which are often related to the military sector and which are mostly financed to a substantial extent by government funds, can only be carried out if the costs and thus the risks are being shared by different nations. Many of these projects are in areas where European efforts would lag far behind those of the United States if they were not synchronized (for example, high-energy physics, aviation, and aerospace and military technology). Apart from this incentive of competition, there are very costly projects or systems that represent a challenging task to satisfy urgent social needs by cost-sharing international cooperation (for example, advanced international public transport systems, automotive safety, and environmental R&D).

The sharing of resources has also to be seen in a more technical sense. The high degree of specialization, both in know-how and equipment, necessary in advanced research and technology has led to a trend to specialization of different firms and even of different countries in certain fields, or at least to a situation where only a few specialized companies in each country have sufficient resources to be engaged in the more advanced and complex interdisciplinary research areas. It is therefore to be expected that large diversified projects will be tackled by combining the special capabilities of different national firms or countries. This may include, to some extent, the recruiting of highly specialized people from different countries for a joint project.

Enlargement of the Market

In many cases of advanced technology, the research and development costs are extremely high in relation to the total turnover expected for the product. Very often these high investments can only be justified if the market is extended through cooperation between firms in different countries; the joint product can then be sold on the enlarged market.

The Need for Standardization

Large systems or complicated technical products can often be made economically only if they are standardized in as many countries as possible. Examples of this are standardized software for the computer industry, standardized modern transport systems (such as the ultra-rapid trains being studied at present in different countries), and standardized telecommunication systems. The standardization will only be achieved through close cooperation among the different national industries during the research and development phase.

Political Reasons

Government policymakers may also have an interest in fostering international technical operations. In this respect, science and technology policy is considered an essential element in foreign policy. Scientific exchange and technological transfer are frequently agreed upon between governments to demonstrate their close relations and to document their intentions to cooperate in certain fields of mutual interests. Such treaties are signed, however, predominantly in fields that do not necessarily have a "commercial bias."

COOPERATION FIELDS OF PARTICULAR INTEREST TO INDUSTRY(11)

Technical cooperation on an international scale is a relative newcomer, but one that is firmly established on a substantial and growing scale. Besides the political motivation for such cooperation, there are other major incentives: the size of major projects, such as in energy and aerospace; the growing awareness of public service and environmental needs; and the growth of global technologies, such as meteorology and oceanography.

Such cooperation is more difficult than that between companies alone because of the very different roles, motivations, and perspectives of industry and government. It is important, therefore, that the projects chosen require this most difficult form of cooperation in order that they may be successful.

Government and industry are two very different partners, each of whom has unique contributions to offer. Government can offer the ability to cooperate with other governments, as well as financial assistance. Industry in turn has flexibility, management skills, and technical judgement. As a corollary, the motives for entering into cooperation are also different for the two partners: with governments, the motive is predominantly political; with industries, the motive is commercial.

What kinds of projects lend themselves to a multi-national approach with government participation? Clearly, projects which can be carried out and are executed without government intervention do not qualify. On the other hand, as has been mentioned repeatedly, there is high priority on non-market-oriented projects, those projects where the social interest is higher than the industrial interest, as in education, environmental control, and ground transportation.

It has been pointed out that government decision mechanisms are slow — slower certainly than industrial ones. The planning stage of such projects is thus long; accordingly, short-term projects cannot qualify.

Project Selection

Existing means of consultation between governments and supranational bodies, on the one hand, and industry, on the other, fall far short of the ideal when selecting projects for cooperative technical agreements. The European Economic Community (EEC) is farthest advanced with proposals for cooperative effort. Even with these EEC proposals, however, there is a strong feeling that they have not been chosen as a result of a strategic study of the needs of the community and their priorities; nor has there been adequate organized industrial input to the policy formation. There seems to be a need for a point at which social needs can be brought together with a realistic view of what opportunities industrial technology could provide to meet these needs.

A systematic study of the needs in transport, communications, education, environment, conservation of resources, and so forth could serve to isolate those areas which are not sufficiently market oriented for the normal economic forces to operate and where participation of governments would be essential to set communal priorities and sponsor R&D.

More information is needed on the mechanisms of consultation in starting international technical cooperation between industry, governments, and supranational bodies, and on a basis for the systematic strategic studies to identify priority areas for government-industry action.

EVALUATION OF DIFFERENT FORMS OF INTERNATIONAL SCIENTIFIC AND TECHNOLOGICAL COOPERATION AGREEMENTS

There seems to be a widespread feeling of unhappiness, of malaise, about international cooperation in general. No one, beyond making general official statements, has made an effort to analyze the existing pattern and to elaborate recommendations for improvements. At present, no authoritative evaluation exists. If such an effort would be undertaken, no doubt the results would be different for governments and for the private sector.

As far as the United Nations System is concerned, the General Assembly has recently requested the secretary-general to prepare a basic study of the activities, mandates, and working methods of various members of the United Nations System in the field of science and technology for development, to examine the possibilities of improving the efficiency of the system in that field. This report will be available late in 1981. For obvious reasons, this interesting study will be part of an internal review process, a sort of self-evaluation prepared by experts within the secretariats of the various agencies. Such an assessment carried out by independent government authorities or by other independent experts has yet to be commissioned. Since intergovernmental organizations are subject to a decision-making process which is more dependent on political considerations than on more measurable scientific, technological, or commercial criteria, it would be very difficult to reach agreement on the objectivity of any proposed assessment criteria.

The situation in the private sector is quite different. To evaluate the effectiveness of international cooperation agreements in science and technology, no questionnaires are necessary. The commercial results decide in most cases on continuation or termination of such arrangements.

HOW TO "USE" THE INTERNATIONAL COOPERATION SYSTEM

To make improved use of the international cooperation system, an obvious precondition is a better knowledge of the actors as well as the rules of the game. It is astonishing to what extent, even for generally well informed large corpofations, the system of international organizations has remained to a large extent "terra incognita." Both the private sector and the intergovernmental organizations could greatly benefit from a better understanding of the cooperation opportunities which each could offer to the other. I am not so much referring to the lobby function for the protection of the interests of the private sector; this is usually well done through organizations like the International Chamber of Commerce or the Business and Industry Advisory Committee (BIAC) to OECD or through the numerous professional associations which cover practically every industrial branch. I am more concerned about the wealth of information and about direct cooperation opportunities which companies could directly derive from the international system.

There is, first, the permanent statistical data collection which almost every international organization generates on its field of competence. Demographic trends on a country-by-country basis, as well as the latest data on energy, minerals, commodities trade, employment patterns, transportation, or any conceivable other aspect of international relations, is available from the international system.

Second there are special studies on emerging new issues of concern to groups of countries or to all countries of the world, such as: the impact of microelectronics on growth, industrial structures, and employment; robotics; the liberalization of captive markets in the field of telecommunications; privacy and the free flow of information; technology assessment; and transnational corporations in world development.   A simple request for inclusion on the mailing list to the publication offices of the various international agencies concerned would be a first step to be informed on a regular basis.

Third, international agencies like UNESCO, ECE, and OECD provide detailed country profiles on the organization of the national science and technology system of member states. Furthermore, to prepare for the 1979 United Nations Conference on Science and Technology for Development, more than 140 national position papers were prepared, which provide a collection of unique data on the profile of national science and technology infrastructures as well as of the aspirations of practically all countries of the world.

Fourth, the international funding institutions like the World Bank, the Regional Development Banks, and the United Nations Development Program (UNDP) are permanently screening the national development plans of most countries in the world, and thus provide a wealth of macroeconomic and microeconomic information for cooperation purposes which are otherwise not always easily obtainable.

Fifth, more personal contacts can be established and personal experience can be gained in making technical experts available by participating in advisory panels and by assisting, usually in an observer or expert function, at international conferences (such as UNCSTD, Law of the Sea, UNCTAD, UNIDO, and Conference on New and Renewable Energy Sources).(12)

Sixth, there is a general trend by the administration of donor countries to have the private sector more involved in the execution of technical assistance programs. The potential for such cooperation is enormous. According to an United Nations assessment,(13) excluding the activities financed by the World Bank, the United Nations official developmental assistance in 1978 was $2.2 billion, while the total sum of development assistance from all sources excluding centrally planned economies was about $21 billion. It is estimated that the resources committed to endogenous capacity in science and technology are of the order of 8 to 10 percent. Of course, the involvement of the private sector of the given donor country is not automatic, but needs the agreement of the respective recipient country. Nevertheless, there is sufficient room to increase for reasons of mutual benefit the scope of technical cooperation agreements with all interested groups in both parts of the world in the public sector and the private sector, as well as at academic institutions.

Seventh, in this connection, it is worthwhile to mention the important role of consulting engineering design organizations as agents for international cooperation in the field of science and technology. It is generally little known that, for example, in the case of the World Bank, an estimated 25 percent of the bank projects are executed through consulting engineering organizations in developing countries.

SUMMARY

International scientific and technical cooperation is becoming a rapidly increasing important factor of international relations, for governments and for the private sector.

At the same time, the mechanisms of such cooperation are surprisingly little understood.

One of the explanations for this unsatisfactory situation is that anything called "international" or "multi-national" seems to be somewhat suspect and, in any event, out of reach for the normal daily concerns of individuals who, after all, live in a predominantly national context.

Any political or economic crisis situation tends to favor a more conservative national approach – or as some argue, a more egotistic-nationalistic approach – to international affairs.

Any international cooperation is by its very nature a mixture of political interest on one side and commercial or scientific interests on the other.

As a general rule, one could argue that smaller countries tend to get, in comparison with their input, a better deal out of their membership in international organizations than do larger countries. In the private sector, just the opposite seems to be true: because of the complexity of international relations and the high number of actors involved, it is almost impossible for a small or even medium-sized company to be able to assess and digest this potential adequately. The larger and mostly multinational operating companies are well advised to institutionalize an analysis unit specializing in data from the international system.

International organizations, governmental or nongovernmental, are as good or as bad as their members wish and allow them to be. This conference has, therefore, already served a purpose if it has demonstrated the positive commercial and technical opportunity potential of the international system.

NOTES

(1) Organisation de Cooperation et de Developpement Economiques, Organisations Scientifiques Internationales (Paris, 1965), pp. 11-31.

(2) United Nations Economic and Social Council, Economic Commission for Europe (ECE), Analysis of Institutions and Procedures Relating to the Management and Organizationn of Co-operative International Research, SC. TECH/R.41/Rev.1 (Geneva, May 31, 1977).

(3) See also Chapter VIII.

(4) ECE, Analysis of Institutions, paragraphs 13-17.

(5) See, for example, the following:
United Nations General Assembly, Overview of Activities of Organs, Organizations and Programs of the United Nations System on the Field of Science and Technology, A/CONF.81/PC/19 (Part II) (New York, April 16, 1979).
Union of International Associations, Yearbook of International Organizations (Brussels, 1977).
Eugene Skolnikoff, "The International Imperatives of Technology," In Technological Development and the International Political System, (Berkeley: 1972).
M. Elmandjra, The United Nations System: An Analysis (London: 1973).
E. Haas, M.P. Williams, and D. Babai, "Scientists and World Order," in The Uses of Technical Knowledge in International Organizations, (Berkeley: 1979).
Commission of the European Communities, Prospects for Co-operation Industrial Research in the European Economic Community (Luxemburg, 1973).
Council for Mutual Economic Assistance, Co-operation of the CMEA Member Countries in the Field of Science and Technology (Moscow, 1976).
Council for Mutual Economic Assistance, Information on Some Nine Nations Concerning the Activities of the Bodies of the CMEA in the Field of Scientific and Technological Cooperation (Moscow, 1975).
J. Gueron, "Aspetti Metologici e Organizzativi della Cooperazione Scientifica e Techologica in Europa," Scienza e Tecnica (Milan, 1973), pp. 577-586.

(6) EIRMA, Co-operative International Research, Working Group Report No. 9 (Paris, 1972).

(7) This chapter synthesizes a much more detailed approach presented in United Nations Economic and Security Council, Economic Commission for Europe (ECE), Analysis of Institutions, paragraphs 54-139.
See also F. Koenigs, "Antitrust Implications of International Cooperation in Research," "Main Speeches," and "International Co-operation in Research and Technology," in EIRMA Conference Papers, vol. 10 (Paris, 1971), pp. 55-69; and K-H. Standke, Europaische Forschungspolitik im Wettbewerb (Baden-Baden: 1970), pp. 145-193.

(8) EIRMA, Co-operative International Research.

(9) EIRMA, Co-operative International Research, pp. 13-14; Also, K-H Standke "La Cooperazione Tecnologica Europea nel Settore Privato," Scienza e Tecnica (Milan), 1973, pp. 595-600.

(10)   See A. McKnight, "La Cooperazione Tecnologica Europea nel Settore Publico," Scienza e Tecnica (Milan), 1973, pp. 589-594.

(11)   EIRMA, Co-operative International Research, p. 22; see also A.P. Speiser, "Summing Up," "Main Speeches," and "International Co-operation in Research and Technology," in Eirma Conference Papers, vol. 10, p. 95.

(12)   The World Bank, Uses of Consultants by the World Bank and Its Borrowers (Washington, D.C., 1974).
        See also, The International Advisor: His Role, His Selection and His Training, (Univerisity of Ottawa Press, 1974).

(13)   United Nations General Assembly, A/35/224, Table 3.

LIST OF ABBREVIATIONS

AAAS	African Association for the Advancement of Science
ASEAN	Association of South East Asian Nations
ASCA	Association for Science Co-operation in Asia
BIAC	Business and Industry Advisory Committee (to the OECD)
CERN	European Organization for Nuclear Research
CISTOD	Confederation of International S & T Organizations for Development
CMEA	Council for Mutual Economic Assistance
COST	European Cooperation in the Field of Scientific and Technical Research
ECE	Economic Commission for Europe
ECSC	European Coal and Steel Community
EEC	European Economic Community
EIRMA	European Industrial Research Management Association
EMBL	European Molecular Biology Laboratory
ESA	European Space Agency
EURATOM	European Atomic Energy Community
FAO	Food and Agriculture Organization of the United Nations
HABITAT	United Nations Centre for Human Settlements
IAEA	International Atomic Energy Agency
ICSU	International Council of Scientific Unions
IEA	International Energy Agency (OECD)
IFIAS	International Federation of Institutes for Advanced Studies
IGCP	International Geological Correlation Program
IGO	Intergovernmental Organization
IHP	International Hydrological Program
IIASA	International Institute for Applied Systems Analysis
IIED	International Institute for Environment and Development
IOC	International Oceanographic Commission
IRI	Industrial Research Institute
ISO	International Organization for Standardization

ITU	International Telecommunications Union
MAB	Man and Biosphere Program
NATO	North Atlantic Treaty Organization
NGO	Nongovernmental Organization
OAS	Organization of American States
OAU	Organization for African Unity
OECD	Organization for Economic Co-operation and Development
PREST	Scientific Research and Technical Policy Group (European Communities)
UNCSTD	United Nations Conference on Science and Technology for Development
UNCTAD	United Nations Conference on Trade and Development
UNDP	United Nations Development Programme
UNEP	United Nations Environment Programme
UNESCO	United Nations Educational, Scientific and Cultural Organization
UNIDO	United Nations Industrial Development Organization
UNU	United Nations University
WAITRO	World Association of Industrial Technological Research Organizations
WFEO	World Federation of Engineering Organizations
WFSW	World Federation of Scientific Workers
WHO	World Health Organization
WMO	World Meteorological Organization

5
Industry's Role in International Technical Cooperation
J.E. Goldman

In assessing the role industry can play in international technical cooperation agreements and, more particularly, how industry can benefit, it is important to place in perspective several underlying assumptions.

1. Governments have no technology to exchange. In the market economy countries, nearly all of the appropriate technology belongs to the private sector, is paid for largely by them, and is often a critical asset in a company's balance sheet. (The term "appropriate" used here excludes those technologies developed for military applications or in pursuit of national goals and which are paid for by governments.) To the extent that international technical cooperation is to be considered an instrument of political objectives, there will be finite costs associated with the commitment of technology that normally lies in the private domain, and these must be borne by somebody. These costs are not likely to be assumed by the private sector unless there is a real or forseeable quid pro quo. This implies a need for governments to learn to work with industry and seek adequate means to recompense industry for the utilization of their justly begotten assets for purposes other than the accepted goals of the private sector, namely, sustained profitability.

2. Dr. Hawkins, in Chapter 3, represents that cooperation or the need thereof is a consequence of the existence of national boundaries. As he put it, without national boundaries, it is simply a question of technology transfer such as from Hamburg to Mannheim. I would suggest that this is an oversimplification; that transfer of technology from Silicon Valley to the U.S. Midwest has much in common with efforts to transfer technology across national boundaries; that, in fact, efforts by Kansas City, Philadelphia, and others to develop their local equivalents to Silicon Valley or Route 128 to enhance local growth by the use of the newest technologies are not unlike the EEC efforts to develop an indigenous microelectronics capability that can be shared by all the members of the Common Market.

3. In the new postindustrial society which we are entering, in which the currency of technological exchange becomes more software oriented than hardware oriented, technological cooperation and transfer assumes a new dimension. Software can become a great equalizer among nations, minimizing the role played by natural resources and capital expenditures in easing the flow of technology.

With these three constructs as a backdrop, let us look at some clues as to where opportunities for cooperation exist, particularly as they affect industry.

Optimum cooperative arrangements are intra- and inter-company. Here needs are easily definable, goals are easily established, each partner recognizes objectives when they are achieved, and each jurisdiction benefits in the long run through the internalization of the new technologies and the creation of a cadre of trained and informed people.

It is generally recognized that few if any artificial stimuli are needed to smooth the flow of science across national boundaries. Science has its international system of publications, conferences, seminars, cross-visits and, most particularly, the colleague network that has grown up in the scientific community, which is the most effective proven means of information transfer; and science, after all, is information.

Technology, on the other hand, is not so blessed. Its breeding ground is largely industry, which tends to be insular, dominated by proprietary concerns and shrouded in secrecy. What it needs is a colleague network to rival that of the scientists – not to circumvent the recognized needs for security where required, but to provide a ready channel for communication. Cooperative arrangements between companies are an ideal catalyst to further such contacts. Dr. Hawkins already mentioned some of these. They are particularly appropriate where each partner and each jurisdiction has some complementary resource to contribute. One area, however, that has not yet been fully exploited by the industrial community is that of cooperative relationships with universities across international boundaries. To my knowledge, only IBM has pursued this option with agreeable results.

There are a few international programs or institutions that have proven reasonably successful in fostering cooperative arrangements between companies in different countries and which bear emulation. One is the International Institute for Applied Systems Analysis (IIASA) in Laxenburg, Austria, one of the more successful experiments in international cooperation in research and probably the only one involving East-West cooperation. IIASA has recently instituted an industrial associates program whose near-term goal is to bring industry into the IIASA fold for both financial support and program participation. In the long term, it should constitute an institutional platform where representatives of diverse companies and diverse industries can work together.

A second example is the U.S.-Israel Binational Research and Development Foundation (BIRD-F). This is a foundation endowed jointly by the U.S. and Israeli governments to support industrial research by cooperating U.S. and Israeli companies. A prerequisite for program funding by the foundation is an agreement between one U.S. and one Israeli company to conduct joint R&D. In the approximately three years of active existence, it has proved eminently successful in promoting corporate cooperative arrangements on an international scale.

Finally, there is the example of the Battelle/Korea agreement which led to the establishment of the Korean Institute for Science and Technology (KIST) cut more or less in the image of Battelle. This is more of a research-institute-to-research-institute form of cooperation than an inter-company form of cooperation, but since the preponderance of the clients of both institutions are corporate, it is indeed a form of international industrial cooperation.

II
Strengthening the Scientific Base

6
Materials
Roland W. Schmitt

In this chapter we are to be especially attentive to opportunities to strengthen the scientific base of materials technology for industrial applications. Even with the limitation to materials, the subject is quite broad and challenging. In a sense, we still have to deal with the entire range of issues of this conference — examining materials science simply as an example of the broader issues.

My thoughts come from the perspective of a large, publicly owned U.S. industrial firm heavily involved in international commerce.* This chapter is not a research paper; it is rather a set of observations, comments, and remarks based on personal experience and perspective.

To begin our consideration of international technical agreements, let us review the objectives of each sector involved. First, governmental objectives can include the following:

Improving international relations and supporting foreign policies. Programs aiming at this objective tend to have great symbolic value and often achieve their primary intent.

Promoting trade and economic development through the establishment of standards, the acquisition of foreign technology for domestic use, and the early introduction of sellers to potential markets for the development of reciprocal trade. Programs so motivated have a direct, important impact on industrial activity and should have industrial involvement at all stages.

Addressing technical problems which extend beyond national boundaries. This includes acid rain, pollution of ocean and waterways, aerosols, weather forecasting, and so forth. Industry has a large stake in the outcome of international agreements in these areas, but the degree of direct industrial involvement will be dependent upon the particular subject involved. I shall come back to this later.

---

*I am indebted to Dr. Charles M. Huggins, Manager-International Programs, GE Corporate Research and Development, for invaluable help and ideas in formulating these remarks.

Fostering basic science. This is a legitimate and important function of every government and one of considerable long-range importance to industry. The industrial sector should not only encourage and aid government-to-government activity in this area, but should participate whenever it can support or foster the undertaking.

There are other possible objectives of government-to-government agreements, such as sharing the cost of expensive projects. However, in terms of costs alone, most such projects could really be carried out unilaterally by any of the major governments involved. The primary value is probably the exchange of ideas by scientists from different nations with different perspectives. Also, the development and improvement of technology for national security and defense could be the basis for government-to-government agreements. However, this area is so dominated by secrecy and unilateral national interests and concerns that it is difficult to say anything about it in the context of this conference.

INDUSTRY'S ROLE

Let's turn now to the industrial sector. When we speak of the "private" or "industrial" sector, particularly in relation to agreements by governments, we are dealing with an exceedingly complex array of relationships. The distinction between private and public, governmental and nongovernmental varies from nation to nation, as does the degree to which economies are centrally planned or market controlled. I am going to finesse the issue of public, private, or state ownership by concentrating on the industrial sector independent of ownership. Members of this industrial sector would include such varied organizations as General Electric, Philips, Siemens, Hitachi, the Ministry of Electrical Engineering Industries in the USSR, and the government-owned National Coal Board in England.

With this in mind, I believe that essentially all of the industrial sector's objectives can be derived from one very simple one: to acquire technology of commercial value more cheaply and quickly than through unilateral internal development. Although the removal of the uncertainty of development may seem in theory to put a slight premium on the cost of acquisition as compared to the cost of internal development, in fact, those nations and firms who have been most successful in the past in technology acquisition have seldom had to pay this expected premium over initial costs of development.

Instead, the initial developer, with costs already sunk, has priced his technology and know-how relatively low just to obtain current incremental income. Nevertheless, this objective of the industrial sector is congruent with the governmental objectives of promoting trade and economic development, and thus it is in this area where government-industry cooperation can be most effective.

It is a primary thesis of my points concerning the materials field (and, I suspect, in others also) that the best strategy for governments wishing to foster international trade and economic development is to encourage direct industry-to-industry agreements rather than to become themselves the lead agency for such agreements. Governments can play a role in promoting and encouraging such activities, but the most fruitful and productive arrangements are certainly going to come from the interested industrial parties negotiating directly with one another. My own company has a large number of technical cooperation agreements with industrial concerns throughout the world – in Japan, Western Europe, England, the Soviet Union, and the Third World. In only one or two instances do these relationships depend operationally upon overlying government-to-government agreements, but all of them depend critically on the role of government in establishing the legal and political framework for those relationships.

MATERIALS TECHNOLOGY: SUBFIELDS AND COOPERATION

We first need to identify the technical subfields that are receiving the greatest attention today or that look most promising for the future from the perspective of our conference. Instead of the conventional categories such as metals, ceramics, plastics, magnetic, semiconducting, and so on, I suggest it is more productive to look at the subfields below.

Materials Processing. These technologies, whether pertaining to metals, plastics, glass, ceramics, or others, have become increasingly important to industry because of their impact on industrial productivity and on the efficient use of resources such as capital and energy.

Information, Sensing, and Controls. This area is not conventionally thought of as part of materials science and technology. But it is of immense industrial importance today because of the strong trend toward higher and higher degrees of automation. These technologies are closely related to materials processing technologies.

Materials Substitution. Because of the changing relative costs of energy versus materials, or of one material versus another, materials substitution technologies are growing in importance. Industry has large opportunities to respond to changing conditions of economics and resource availability through the invention of new materials to be used as substitutes for ones currently used.

Resource Recovery and Conservation. Again, the escalating costs of many raw materials and energy are creating the need for new technology in this area.

Environmental Effects. The environmental and health impact of the production, processing, and use of materials is growing more and more significant in the developed countries and is beginning to emerge as an issue in developing nations.

International Standardization. This is a field of obvious importance for international trade and is an area of continuing government involvement.

Among these subfields of materials technology, the opportunities for and value of international agreements vary greatly. In the areas of materials processing, automation, and materials substitution, the stimulus of free markets and competitive advantage provides sufficient incentive to the industrial sector to undertake the necessary R&D. And these industrial organizations may sometimes find it advantageous to consummate agreements with their counterparts in other countries. Obviously the laws, policies, and attitudes of national governments can aid and abet or can hinder and obstruct such efforts, but the direct intervention of governments is really not needed if natural economic forces are allowed to operate. Thus, although no government-to-government agreements are required in these areas, it is desirable for governments to assure that obstacles are removed from the paths of industrial organizations who may wish to establish international industry-to-industry agreements.

In the areas of resource recovery and of conservation and environmental effects, market and competitive forces may not be strong enough to stimulate the desired degree of international cooperation. Because many of the problems in these areas are common to all industrial organizations engaged in international trade, and because there may not be competitive advantage in solving such problems, these areas might be helped by the initiative and leadership of government-to-government agreements. However, as governments proceed in these areas, they should involve industrial organizations – possibly through trade associations – in the initial planning and discussion stages.

In the area of international standardization, it is my impression that direct interactions between the standardizarion agencies in the various countries are working quite satisfactorily.

There are already large numbers of government-to-government agreements pertaining to science and technology. A number of them pertain to the field of materials. For example, under the U.S./USSR cooperative agreements of 1972, there is an international agreement on metallurgy, and a task force on metallurgy was established; the U.S./Israel Binational Science Foundation Agreement of 1972 has provisions for basic and applied research on alloys; and there are many more, including other examples of government-to-government agreements in the field of materials between governments other than the United States.

VALUE OF INDUSTRIAL PARTICIPATION

There is little evidence of U.S. industry involvement in these agreements except that a number of industrial scientists have been included as members of scientific exchange teams. Their role is generally that of individuals rather than representatives of industry. The principal motivation for these agreements has been the use of science and technology as an instrument of foreign policy. I am in no position to judge whether or not these agreements have been successful in achieving that objective. Others must make that judgment. But from my own observations and from anecodotal evidence, I must conclude that there have been few, if any, industrial benefits that have come from these agreements so far. This is not meant to say that they have failed in their objectives. In fact, there is no evidence that these agreements were targeted primarily at industrial interests in the first place.

But it does raise a question about the premise of this conference. That premise contends that government-to-government technology agreements will be made in any event and that we should therefore tap into them for industrial spin-offs even after the fact. But such an approach will always produce results of marginal value to industry. There are several reasons that mitigate against obtaining maximum industrial benefits from government-to-government agreements that are conceived and consummated with purely political objectives in mind and without an initial, deep concern for the industrial value and implications of the agreement.

First, most industrial R&D is done to obtain proprietary or competitive advantage in the marketplace. Unless this objective is kept in mind at the time technical agreements are formulated, it is unlikely that the agreement will deal with the intricacies inherent in the competitive milieu. Second, much industrial R&D is too fast-moving to be tied to the slow-moving pace of government-sponsored international programs. Third, industrial programs generally require much more definite goals than are specified in most government programs, and the industrial parties need to retain the right to change those goals in midstream. Fourth, much industrial R&D has a fairly large engineering content, and a mutually productive two-way flow in engineering is more difficult to structure than such a flow in basic sciences. Now all these difficulties can be handled adequately in technical agreements if the potential industrial rewards of such an agreement exist and if these factors are kept in mind at the time that the agreements are formulated. Obviously this is seldom the case, when foreign policy has been the principal driving force behind these agreements.

There are other constraints that must be dealt with if productive industry-to-industry cooperation is to be achieved in the materials field. The industrial and competitive structure in some materials businesses (for example, plastics, pharmaceuticals, and diamonds) is so intense as to mitigate against any type of agreement other than, perhaps, patent-licensing exchanges. Also, especially in the United States, the antitrust laws and export-control regulations place severe constraints on the negotiator. Export-control regulation is undergoing a major change in the United States today that may have a profound impact on the international flow of scientific and technological information. A shift is occurring from an emphasis on control of exported products to control of exported technologies. It is too early to tell exactly how these changes will impact the international exchange of science and technology. In my view, though, there is the potential of a great threat to international cooperation. It should be a matter of prime interest to all of us to follow these developments closely and to ensure that there is still adequate opportunity for beneficial international exchanges of science and technology.

BENEFITS AND CONSTRAINTS

My conclusions and recommendations are fairly simple. First, if industrial benefits are to be gained from international technical agreements in the materials area, the industrial objectives should be kept in mind from the beginning of the negotiations and from the beginning of the conceptual formulation of the agreement. Thus, the first requirement is to understand and agree upon the real purposes of the agreement. Second, the delegations that negotiate such agreements should be staffed with people who understand and can deal with complex industrial issues that may be beyond the knowledge and experience of the experts in intergovernmental affairs. Third, it is important to understand and to consider the limitations and barriers to industrial sector participation, especially so as to avoid misleading the other parties in the negotiations who may have quite different constraints. Some of the factors that must be uppermost in the minds of U.S. negotiators are export-control regulations, antitrust law, proprietary interests, patent rights, economic and business benefits, and cost and budgetary factors. Fourth, where private sector participation is desirable, delegations should include people who have had some experience in managing such cooperative programs. Fifth, it would be highly desirable for the governments involved to see that their representatives brought with them a truly integrated set of governmental policies. In the United States, for example, the government itself should ensure the integration of the policies of our Commerce, State, and Defense Departments and should not require private industry to make this integration.

In conclusion, there are a few enduring facts that should be kept in mind. The scientific and technical communities have traditionally been international in scope, and most scientists and a great many engineers regard themselves as part of this international community. To them, the merit of a discovery or invention is not confined by national boundaries. Today, also, larger and larger proportions of individuals engaged in business and commerce are beginning to think in international terms. Both the sources of technology and the markets for technically based products and services are becoming more and more international in scope. Thus, I believe that the primary strategy of governments in pursuing international technological agreements should be to foster, enhance, and utilize these natural trends and tendencies in the scientific, technical, business, and commercial worlds. If we try to start from the premise of a mutuality of interests, and structure agreements to fit these mutual interests, I believe that the outcome will be much more productive than a strategy of simply trying to exploit agreements based on rather narrow premises after they have already been negotiated.

7

Overview of Policy Issues: Panel Report

Charles Crussard

Materials are strategic to the industrial base of modern technological society. It has been estimated that annual consumption of minerals in the United States alone is 40,000 pounds per person. In general, issues in the materials field are viewed through a prism consisting of interrelationships of the economy, energy, and the environment. This is reflected in an array of policies affecting the cycle of materials from extraction to waste disposal, within a country and internationally. More specifically, attention among the industrialized countries is focused increasingly on availability and access to critical materials and on enhanced productivity in their processing.

The considerations of this panel, therefore, reflected major concerns in the materials field within this context. The two questions that guided the considerations were:

1. Are international technical agreements appropriate vehicles for improving productivity and innovation in the materials industry?
2. If so, in which areas and under what circumstances is such cooperation most beneficial?

The members of the group presented various cases of international agreements, successful and unsuccessful. There are many cases of successful bilateral agreements initiated by private industry partners, frequently between a producer and a user, but also between two coproducers. These agreements are generally aimed at a given product.

There are cases of successful multilateral and multinational agreements involving either government agencies only (such as the agreement on isotope separation by centrifugation between Germany, the Netherlands, and the U.K.) or private industries and governmental parties (such as the liquid hydrogen powered aeroplane, initiated by Lockheed and involving now three government organizations and four private companies in five countries, or the European COST program on gas turbines). These are generally big and complex projects, too large for one company or even one country.

Other cases of multinational agreements in technical areas have been unsuccessful when organized by government agencies for political reasons and without sufficient evaluation and assessment of the project.

Multinational agreements are very well suited in domains involving such concerns as standardization or qualification.

The panel concluded that the most successful agreements were:

- Bilateral, between two parties in two countries;
- Multilateral, between private companies;
- Multilateral, between private industries and/or government agencies, on projects initiated either by industry or by governments – provided, in the latter case, industry is consulted from the start and during the process of development

The domains where multilateral projects have the best chance of success are the following:

Environment. Large differences in pollution regulation, from country to country, may lead to discrepancies in profitabilities of materials industries in different countries. By reaching agreement on more integrated and consistent regulations, international cooperation may be quite helpful.

Energy saving, mostly in extraction and production of raw materials. Here, the incentive of private enterprise can be regarded as sufficient, and international agreements are not particularly necessary.

Substitution of materials. This is another area where international cooperation evoked interest, but the conditions of working on such a project were not defined.

Recovery and recycling of materials. Considerable research is now focused on reducing wastage of materials in manufacture and use. International collaboration can be useful here, and will be easier and very profitable in knowledge of basic data, methods of evaluation, standards, and so forth, especially in domains not directly related to proprietary processes critical to the industries involved.

Studies on basic properties and frontier research. There do not seem to be insurmountable problems with cooperation in frontier research, since such basic research is not marketable in itself, but rather a building block toward more proprietary work. The use of international technical agreements in this area would be most satisfactory, because all interested parties could share in the costs and resources of the research as well as gain from the expanded reservoir of knowledge. All possible mechanisms could be used in this basic domain: exchanges of data, seminars and conferences, exchanges of research personnel, and cooperative research programs.

The panel proceeded to survey fields where the lack of basic knowledge was most prejudicial to the development of the materials industry. In this respect, the following topics were proposed which could form fruitful areas for cooperation involving the private sector under appropriate conditions:

- Basic physical and thermodynamical properties of materials
- Influence of trace elements on physico-chemical reactions
- Process modeling
- Effects of various substances on the environment and public health
- Used materials and wastes; exact knowledge of their composition and evolution in time
- Hidden mineral deposits; technique for their prospection
- Renewable sources of materials (wood, plants)
- Repertory of laboratories working on basic properties of materials.

8
Chemistry
Jean Cantacuzène

Better knowledge and understanding of the social and technical attitudes of other countries can contribute greatly to improving the industrial productivity of one's own country. This process of communication is facilitated through international cooperation, especially in the technical sphere. Traditionally, government intervention has played the determining role in technical cooperation on an international scale. It is at the government level that one gets a fully comprehensive view and that one can therefore establish guidelines for cooperation that reflect the range of a country's interests, including the improvement of productivity. Conversely, the role of governments should not be to prescribe technical cooperation in detail, but rather to establish the climate and main trends for cooperation.

One observation I would like to make on the theme of this conference is based upon my recent assignment in Washington, D.C. as French scientific attaché. It is my deep conviction that the productivity of a country or an industrial sector is much more affected by major socio-economic phenomena that are known to everyone than by technical information that is restricted and protected by proprietary rights. We can learn a great deal from the broad experience in other countries with factors such as those below which impact productivity.

Government-industry relations in the pollution/regulation cycle have frequently been characterized by more pollution, growing regulation, and an adversarial posture. The experience of the chemical industry in some countries, however, may offer a model for reduced pollution, fewer regulations, and greater agreement on objectives; international cooperation on ecological problems involving the private chemical industry could be helpful, in particular for reaching some agreement on pollution standards.

Work relations between employees and managers in Germany and Japan are often based on consensus vis à vis production goals which results in overall productivity gains. Joint seminars on industrial

66

productivity between U.S., European, and Japanese managers may facilitate the sharing of experience in this area.

Energy moderation demonstrated by reduced consumption in Western Europe without sacrifice to high living standards can serve as an example for other countries.

Policies and practices in research and research management which are shared between countries can offer new insights for greater efficiency and longer-term growth. Educational programs and seminars on R&D management like those organized each summer by the Massachusetts Institute of Technology could be expanded on an international basis and organized each year also in Europe and Japan.

Within this broad consideration, however, I would like to focus more specially on a few illustrations of international scientific and technical cooperation which can involve industry on a practical level. My examples are drawn from programs relevant to the area of chemistry.

One mechanism that presents such an opportunity would be a program promoting long-term assignments of industrial engineers and researchers at national or foreign universities to investigate new fields in greater depth.   This is particularly important at a time when industries are often obliged to reconvert or assimilate new techniques. Within this mechanism, a workshop of international mobility of scientists and engineers is jointly sponsored by the U.S. National Research Council, the European Science Foundation, and the NATO Science Committee. In the field of chemistry, for example, frontier research in enzymatic catalysis, biotechnologies, and photochemistry could be suitable for this type of assignment.

A recent program conducted through the Délégation Générale à la Recherche Scientifique et Technique (DRGST) and the University of Texas at Austin may serve as a model for such. an effort. The French government declared in 1974 that enhanced oil recovery was a primary subject of interest. Since this area is a leading specialty at the University of Texas at Austin, DGRST established a program with the university for two French engineers to spend a year on assignment there, improving their knowledge of the physiochemistry of interfaces.

Another mechanism for cooperation in which the private sector can participate involves seminars jointly sponsored by national research institutes of two or more countries. For example, in 1979, the National Science Foundation and the French National Center for Scientific Research (CNRS) agreed to organize joint seminars (twelve persons from each country) on topical scientific matters, such as theoretical links between heterogeneous and homogeneous catalysis; laser chemistry; and materials science. Experts from other countries were also invited. Industrialists, especially Americans, attended these seminars in a specialist capacity. These exchanges foster bridges between the academic viewpoint (the "technology push") and the industrial one (the "market pull") at an optimum international level.

A third mechanism which seems particularly valuable is sponsorship of a program by two or more government agencies directed to the interests of industry. In 1979 and 1980, the French Agence pour les

Économies d'Energie and the U.S. Department of Energy sponsored joint seminars on energy conservation in industry. Two of these seminars have already been held, one in Paris and the other in Washington. This type of initiative between two governments, directed to industry in two countries and involving a major problem in which economy and technology are directly linked, exemplifies a timely exchange of useful information. This kind of mechanism could afterwards evolve to large-scale research projects promoted by governments in which industry could share costs and participate in the research; coal research such as pollution control or large-scale gasification projects may be an area of possible cooperation.

9

Overview of Policy Issues: Panel Report

Karl-Heinz Büchel

Five broad areas for industry participation in government-to-government agreements within the field of chemistry emerged during the panel's considerations: frontier research, educational programs, ecological impacts, toxicology and public health, and standardized specifications for products and systems. The greatest opportunity for such participation may rest in projects requiring substantial financial commitments in noncompetitive areas of research where cost-sharing arrangements and the pooling of complementary resources might take place.

Of particular concern to the panel members was the structure for industry participation that might evolve in an international program of cooperation. Clearly the distinction between competition and collaboration would have to be recognized both by the subject area of a potential project and by the conditions and incentives for such participation. How would the use of a corporation's resources for an international technical program affect projects ongoing in the company? How would antitrust policies apply? Would tax incentives be involved in a program focused on key problems related to public health and safety that were common to the agreement participants? Industry's role in collaboration will therefore be dependent upon the opportunity to participate in defining the nature and scope of a project and the determination of how government policies might affect such collaboration.

ISSUES FOR CONSIDERATION

Most R&D performed in the chemical industry is an element of the worldwide competition characterizing this industry. Therefore, only selected segments of chemical research are suitable for international cooperation. One area that might be a potential field for cooperation, in addition to those surveyed by the panel, is the development of new raw material resources.

Projects demanding a very high financial commitment should particularly be considered under this aspect. Definitely excluded from those considerations would be direct product research and process development. Both are highly competitive areas where proprietary concerns would preclude collaboration.

It is very difficult to give discrete rules by which such cooperation could be established. The political climate and the economic situation can differ greatly from nation to nation. Clearly, a stable economy, roughly the same level of development, and a common interest between the countries involved will foster joint activities. Equally important is a supportive attitude on the part of governments.

From the industrial standpoint, international cooperation through affiliated companies is clearly the most effective. Joint R&D projects are also common between companies and universities of other countries. However, cooperation between different industrial concerns, if not hindered by competition, is often inhibited by antitrust legislation.

General recommendations for private sector participation in activities conducted through international technical cooperation apply in particular to the chemical industry. The major point of guidance is that economic motivation is always the strongest incentive. Where international R&D programs are deemed appropriate with private sector involvement, government policies will have to be designed to accommodate such cooperation. International educational programs can be a primary vehicle for establishing future cooperative programs.

10
Biology
Lewis Sarett

The discovery and development of new drugs and vaccines for human and animal diseases, and the creation of improved pesticides for crops under cultivation, are among the most significant areas for the application of new biological research techniques that have captured the interest of pharmaceutical scientists in the private sector in recent years. The objective here is to review some of these areas of current interest or future promise and to discuss where and how international governmental institutions are, or should be, involved in seeing that they are properly exploited.

The support of basic biomedical research has been a concern of individual national governments in the developed countries of the world since before World War II. Through its support of symposia, visiting fellowships, and research grants in the United States, the National Institutes of Health (NIH) has helped create and disseminate a base of biomedical knowledge which, when appropriate, has been converted by scientists in the private sector into new drugs and vaccines. Research on recombinant DNA – which offers new methodology for production of such polypeptides as human insulin, human growth hormone, and various members of the interferon family – has blossomed in the academic world recently and has stimulated a number of new commercial ventures. A similar prospect is opening for cell culture in the production of monoclonal (homogeneous) antibodies, which are useful in the near term for diagnosis and later may prove to be valuable in therapy.

The hypothetical hazards attributed to recombinant DNA technology led to early regulation and close international monitoring of procedural guidelines and individual practices in each nation where the research was undertaken. Regulations promulgated in reaction to the supposed dangers that some perceived varied widely in stringency and scope from nation to nation. Harmonization of these regulations would have reduced any competitive advantage secured by research organizations in nations with loose guidelines during the initial wave of speculation about the problems and potential of the new technology.

In practice, however, this disparity in individual national approaches to control of the emerging technology has not seemed to work to the disadvantage of scientists in the United States, where perhaps the most stringent regulations were imposed. At this point there does not appear to be any special merit in establishing additional cooperative research in these new developments at the international level, beyond the traditional means of communication for laboratory workers.

Among other desiderata, the reestablishment of scientific ties with The People's Republic of China opens up new research possibilities for the pharmaceutical industry; specifically, the Department of Health in Peking is a repository of literature on Chinese herbal medicine collected over millenia. This move on the part of U.S. political leaders has thus raised the possibility that China and the rest of the world may benefit from practical evaluation of this unique and ancient lore.

Viral diseases respect no national boundaries, and epidemics frequently are international problems. Through efforts of the World Health Organization (WHO), in cooperation with national health agencies, smallpox has now been eradicated. The Pan American Health Organization, like WHO, is a vehicle for channeling international funding into programs to achieve regional and local health goals. The private sector not only supplies marketed products to meet these needs, but also cooperates with personnel of these international agencies in the clinical testing of new medicines.

Many of the Third World nations are situated in the tropics and are thus plagued by parasitic diseases that are rare in the developed countries of the world. Since the research facilities of the pharmaceutical industry are located for the most part in the latter, the bulk of its resources is spent on developing agents needed in the home markets of the individual pharmaceutical firms. Thus, research logically tends to be aimed at treatment of such familiar disorders as arthritis, bacterial infections, mental disturbances, and cancer, rather than on diseases such as trypanosomiasis, malaria, and leishmaniasis. Aggravating the situation are the inadequate public health standards and subminimal health budgets of many Third World countries. In the United States and Europe, pharmaceutical firms are being made increasingly aware of this need to refocus their research priorities, by political leaders who are mindful of the pressures brought to bear on multinational corporations by such international agencies as the United Nations, in an effort to increase expenditures on tropical diseases.

Given this set of problems, an important role can be seen for the national governments under whose jurisdictions various member firms of the pharmaceutical industry operate. As part of their foreign aid programs, these governments could guarantee to the firms markets for the drugs they discover to be of value in tropical medicine and could provide them to recipient nations as part of a program to aid in development of adequate public health systems. Additional incentives could, of course, be created by the governments of developed nations, either individually or in concert with others.

Similarly, research could be stimulated by multi-lateral agreements aimed at the development of drugs and vaccines for treatment of less common diseases (orphan diseases). A coalition among OECD nations, for example, could agree on incentives for pharmaceutical research on myasthenia gravis, multiple sclerosis, cystic fibrosis, and others.

The clinical testing of new drugs and vaccines is subject to lengthy and detailed regulatory supervision. Although individual nations have created their own individual procedures, the basic requirements are similar and all have grown in magnitude and complexity over the years. Moreover, the regulatory agencies keep in touch with one another, adding, as they see fit, alarms and caveats drawn from clinical experiences elsewhere. It would now be appropriate for the OECD nations, perhaps led by the United States, to jointly explore the possibilities for deregulating the new drug-approval process, simplifying the procedures, and jointly eliminating those measures which have become cumbersome and superfluous.

Most of the regulatory agencies have a history of working cooperatively with industry; it would be expected that such a multinational review of deregulation options would be conducted in that same spirit of cooperation, with continued industry input.

Much of the data forming the basis of approval of new drugs and vaccines is toxicological. First, in animals, the candidate compound is administered over periods of time ranging up to a full life span and at doses that vary from very low to sublethal. Toxic effects are noted and classified on a minutely detailed basis. The interpretation of the data is the work of expert pathologists who sometimes disagree with one another on the significance of the study findings. It is not uncommon for drugs to be approved in one country and forbidden in another entirely on the basis of identical animal toxicity data. Furthermore, in the case of France and Japan, local repetition of toxicological (and pharmaceutical) data is required. A measure for harmonizing the toxicity data evaluations and for elimination of duplication could be undertaken by OECD nations, again in concert with industry.

Safety assessment does not stop in the animal study phase of new drug evaluation. It is the force behind the careful study of candidate products in man during controlled clinical trials prior to approval by regulatory agencies.   Some adverse reactions are too rare to be identified with certainty when a new therapeutic agent is in the preapproval phase, and must be the subject of continued monitoring after the product is on the market and in wide use.   This "post-marketing surveillance" has proved to be a difficult process, although certain European countries, aided by centralized drug-purchasing systems, have done a reasonably satisfactory job of it. An international effort to make better use of systems for reporting and communicating adverse reactions after a drug has been placed on the market could, perhaps, be helpful. Such a multinational effort would have the dual goal of eliminating spurious reports and verifying valid ones.

11
Overview of Policy Issues: Panel Report

Bernard W. Langley
Lois S. Peters

Because mankind's existence and well-being is inextricably related to the subject matter of the biological sciences, biological study is inherently international. It is hard to determine the progress of our understanding of the biological sciences in terms of public and private directives, but much industrial innovation has arisen from the study of these subjects. Biology has given rise to the oldest industries — agriculture, brewing, fishing, forestry, and medicine. During the last fifty years, such specialized, highly profitable research-based industries as pharmaceuticals and agrochemicals have emerged from applied biology.

DISCUSSION

Despite the international nature of the subject, industries based on applied biological research have not been involved extensively in formal international arrangements or agreements. There is a consensus that international cooperation in the biological sciences in pure research, especially academic research, is crucial. The perception is that this cooperation functions best at the working-scientist level through conventional fostering and encouragement. International communication among basic laboratory scientists at conferences and through the scientific literature has been extensive in the past and has been critical to the growth of biologically based industries. Technology transfer and innovation frequently occur more rapidly and efficiently with one-on-one communication between scientists, and this interaction and cooperation should not be bureaucratized or institutionalized.

The panel felt that since universities are primarily responsible for pure or basic research, international cooperation in academic research and providing support at the university level would be vital to industries dependent on innovation stemming from the biological sciences. University research and training is almost entirely funded by governments which in turn play the major role in initiating and formulating international agreements. Therefore, industry should devise a mechanism to enter into the process of international agreements affecting university research and education. Since industry for the most part relies on universities to educate, international industrial cooperation at the university level can help broaden the training of scientists and technicians as well as facilitate development of new areas of technology. In fact, the world is changing so fast, especially with regard to our understanding of biology, that more effort should be made nationally as well as internationally to update and retrain the key scientists concerned.

Some specific issues in biology do require unified international attention, and currently there are several opportunities where it would be particularly helpful for the private sector to involve itself in international technical cooperation. These are in the areas of ecological science, technology transfer to developing countries, and regulation and licensing procedures.

In the field of ecology, international data collection, comparison, and study are needed. There are several groups cooperating in such endeavors. International cooperation in the ecological and environmental sciences and opportunities for private sector involvement are discussed in another chapter of this volume.

Although much international cooperation takes place between companies, when dealing with the developing countries or the eastern block, interactions with the governments are more usual, and therefore it is particularly important for industry to participate in the development of international agreements with these countries. Technology transfer is an area where private companies are the most experienced, and it is important to involve them with the governments' agencies concerned with these programs. More active participation on the part of all parties engaged in technology transfer arrangements would broaden the learning process — in both directions — between those transferring the technology and those receiving the technology.

The United States Agricultural Extension Agency is regarded as a model for fostering the use of scientific principles by low technology users. In fact, the Consultative Group on International Agricultural Research (CGIAR) does employ many of the ideas of the U.S. Agricultural Extension Program. CGIAR now supports research and training activities through eleven centers or programs located in developing nations. The combined annual budget of these centers exceeds $100 million. There are several other international organizations focused on transferring technology to developing nations which industry might take note of. The International Development Research Center (IDRC) established by the Parliament of Canada in 1970 encourages coordination of research and fosters cooperation between developed and developing regions. Their areas of concentration include agriculture and nutritional and health sciences. Since its inception, over $100 million of IDRC support has been provided for 550 research projects of which more than two-thirds were in science and technology. The Swedish Agency for Research Cooperation with Developing Countries (SARCDC) promotes research that helps developing countries achieve greater self-reliance and economic improvement. The United States is also considering an institute for scientific and technological cooperation (ISTC) which will address the issues of transferring science and technology to developing countries. It is designed to be a grant-making as well as coordinating body and could very well benefit from the economic acumen of the private sector. It is not clear that industry has been participating to any great extent in these activities. However, the point should be made that not only will industry enrich its commercial opportunities by participation in such activities, but we should widen our definition of industrial productivity to include improving the quality of life for a wider range of people and not focus only on those areas directly related to hard dollars.

Much of the panel's time was spent on the issues of regulation and licensing. There was a clear recognition of a need and opportunity for international cooperation in developing methods to accelerate testing procedures without jeopardizing safety.

In the earlier days of the current growth of the biological industries, the public was only too happy to have their diseases cured, weeds and insects killed. Then three things occurred. Increased longevity and populations began to stretch resources, especially in food and fuel. Second, those very things that helped one section of the population began to destroy or be viewed as detrimental to the quality of life in another. Third, although most people were grateful for the newer potent drugs and agrochemicals, the side effects of a few of them raised questions of their overall acceptability. The balance between good and harm became a priority topic of public interest, and officials then sought protection with complex and massive registration procedures and environmental regulation for industrial products viewed as potentially biologically harmful. New biological subject areas grew up in toxicology and environmental science, but despite these new subject areas, regulations were often made before a sufficient data base could be developed and put in a realistic context.

These massive registration and regulation procedures served as a technical disincentive. Twenty-five years ago in the United States, it took $5 million and two to four years to develop a new bioproduct or drug; it now takes $60 million and eight to twelve years to develop such a product. The panel felt that international agreements and technical cooperation could be useful in developing our understanding of the basic sciences which underpin toxicology; in that way, more meaningful regulations and registration procedures can be devised. However, they were doubtful of the value of working to the same regulatory yardstick internationally. Differing standards exist in different countries; for example, Searle's Aspartamine is now sold in France and Belgium, but not yet allowed elsewhere. Different cultural, economic, political, and social contexts can be utilized in this sort of research, and these experiences can be compared and used to develop a better data base for future decision making, and it is in this area that international technical cooperation is critical.

An important lesson of the recent past is that much time and money has been spent in defensive research, and deliberation might be minimized if the private sector, in certain cases, took the initiative in developing guidelines and codes in biologically related industries. An opportunity for industry to become involved in such matters is the development of model legislation and guidelines for nutritional elements in food. Such legislation needs no uniform standard, but can be devised in such a way that socioeconomic conditions in different countries can be taken into account. Industrial participation in this case would not only enhance their credibility, but would also aid them in taking advantage of different markets.

International industrial cooperation in the discussion of ethical problems stemming from biological research may go a long way toward speeding up the testing of products and preventing the delay of marketing valuable products while ensuring the general health and safety of the world population. Several opportunities exist for the private sector to take initiatives in newly developing biological industries based on recombinant DNA techniques. The Committee on Genetic Experimentation (COGENE), a committee of the International Council of Scientific Unions, is currently concerned with guidelines for research on recombinant DNA, risk assessment experiments, and benefits and applications of recombinant DNA. Participation of the private sector in such discussions can only be advantageous in the future as regulatory and registration standards evolve.

ISSUES FOR CONSIDERATION

The difficulty of exploring the frontier edge of biology for profit only increases the need for international cooperation in basic sciences. Cooperation in support of basic research gives industry a window on developing technology. If this is done on an international scale, not only are there more opportunities to develop marketable goods and services, but the nature of the opportunities will be more varied because they will depend in part on the climatic, geographical, and cultural diversity among nations.

Many international organizations do support basic biological science, and there are a few international groups which affect industrial biology. One example is the European Molecular Biology Organization (EMBO), an institute which conducts basic research in molecular biology. Another is the European Economic Community (EEC) program in biotechnology.

Agreements between countries form a basis for cooperation in particular situations and circumstances such as highlighting priorities for concentration of mutual effort, utilizing special resources or overcoming constraints of differing social and political systems. Many agreements in the health sciences are made for one of these three reasons, and although they are not directed primarily to industrial productivity, they inevitably give many opportunities where industry can participate and so contribute to its growth.

Since the first formal bilateral collaborative program in health research and related activities was signed with Japan in 1965, there have been more than twenty bilateral health agreements signed directly between officials of the United States Department of Health and Human Services – formerly Health, Education and Welfare (HEW) – and officials of foreign ministries of health. In the United States, the National Institutes of Health have primary responsibility for conducting health research activities. Although the major extramural activities of the National Institutes of Health are with academic science, even within the national arena, NIH administrators have indicated that the participation of industrial scientists would be welcome. A greater number of industrial scientists should take an active part in international workshops and in visiting scientist exchanges.

The National Cancer Institute is particularly active in international collaboration. Information derived from cooperative programs with foreign nationals contributes to the development and testing of new potentially effective anticancer drugs, and international evaluation of treatment protocols extends the data base for therapeutic approaches and provides new leads for the design of improved treatment strategies. Such knowledge, although certainly used by industrial science, could be more quickly advanced by more active industrial participation.

12
Electronics
Michiyuki Uenohara

Electronics remains the most innovative and technologically inventive industry. Even in today's sluggish state of technological and economic activities, the productivity of the industry is increasing at a rapid pace, while product cost is decreasing despite strong inflationary pressures from rising energy and raw material costs.

Modern electronics has made great progress, owing largely to advances in solid-state-device technologies, especially semiconductor technologies. Semiconductor integrated-circuit technology is rapidly moving into very large scale integration, now widely known as VLSI. This technology will drive the microprocessor and memory to new performance and cost levels, and these devices will extend themselves into many hitherto unimagined areas of application in almost all industrial sectors. The new word "mechatronics," suggesting the marriage of machines and electronics, is rapidly gaining popularity. This and such other new terminology as "informatics," "telematics," and "C & C" (Computers and Communications) indicate the new technologies and markets being created by computers and communications coupled or merged together by microelectronics as a middleman. The waves of innovation in microelectronics are propagating powerfully into every sector of industry. Hence, the semiconductor industry is becoming or already has become a modern basic core industry like the steel industry or petroleum industry.

The cost of a transistor element in modern LSI is about one-millionth of the cost of the transistor twenty years ago, while the cost of textiles has increased by a factor of only 1,000 in the past 200 years. The difference in productivities of these basic industries is due mainly to the difference in their utilization of science and technology, which is clearly reflected in the resources and efforts invested in research and development. Hence, research and development are important not only for the creation of revolutionary new technologies, but also for the improvement of productivity and the reduction of product and service costs. The future of electronics, both in technological development and market innovation, is very bright, especially in microelectronics and optoelectronics and their applications. Therefore, we have to look into every phase of international cooperation in order to make this bright future a reality.

PRODUCTIVITY AND COOPERATION

Opportunities to enhance the productivity of the R&D process, and thus of industrial productivity, through multilateral cooperation involving the private sector differ among the many technical fields and especially among industrial sectors. Gijutsu Doyukai (Japan Committee for Technological Management) has studied the subject of international cooperation for advancing science and technology. The study committee chose energy and electronics as representtive technical areas.

These two have vastly different characteristics in technology, industry, and potential scope for international cooperative agreements. While energy is fundamental to human activity and the world is facing a great energy crisis, all promising new energy technologies are still premature and will require huge R&D investments for many decades, investments which no single company or nation can effectively manage. On the other hand, electronics, as I have already suggested, is today's most innovative and technically inventive industry. The industry is extremely competitive, with every company battling furiously in domestic as well as international markets. Even though electronics technologies such as supercomputers and VLSI do require capital-intensive research and development, there is still ample opportunity for even a lone engineer or a small company to contribute innovative new technologies and products. Therefore, the exploration into the feasibility of greater interaction between the industrial research community and the activities of international cooperative agreements among governments in the electronics field has yielded very discouraging results. An added difficulty arises from the fact that modern electronics is heavily related to military technologies. The Japanese Government has invested negligible amounts of R&D funds in defense technologies, and the industry has either transferred civilian technology to defense or has imported military technology from other countries. Recently, we have been experiencing increased difficulty in importing military-related high electronics technology.

After intensive discussions, the electronics working group of the International Cooperation Study Committee has drawn up the following three basic recommendations:

1. The Japanese balance of trade in electronics products is largely favorable, while the overall technical trade in technology-intensive products is largely in deficit. Japan has to expand its basic research activities and promote the development of innovative technologies in order to improve its technical trade balance. Hence, Japan can contribute to international technological cooperation.

2. Japan has largely followed the original ideas of other nations to advance systems technologies. In order to cooperate, and hence to contribute to the development of new systems of international scale, Japan has to develop an equal level of systems capability. The government has to take more initiative in the development of large systems of national scale, such as space, oceanic, and national resource management and many other systems.

3. For advancing international cooperation, various communications channels have to be enriched in order to improve mutual understanding and trust. The differences in languages, cultures, and physical distances still inhibit fruitful communications. To minimize these international barriers, a national committee has to be organized to explore realistic solutions.

The national committee has not been formed as yet, but many groups from government and industry have been studying various proposals. The results of this conference will be very valuable to their studies.

COOPERATIVE VLSI RESEARCH PROJECT

The cooperative VLSI research project in Japan is an example of coopeative research and development among furiously competing computer makers. This was the very first such experience in Japan, and, if my understanding is correct, it was the first in the world. Managing the project was extremely difficult, but we have gained valuable experience.

Unlike their counterparts in the United States and Europe, Japanese engineers are very low in intercompany mobility. They seldom change jobs at all during their professional careers, so technology transfer due to the movement of engineers is very small. Firms compete strongly against each other and carry almost identical research projects. The efforts of the limited numbers of research engineers in each company are subdivided into many subjects, making it difficult to pose a real challenge on the frontier of technology. If this situation of poor technology transfer continues and risk-taking R&D activity does not increase, Japanese industry may not be able to contribute in high technology areas and may not be able to attain its competitive technological potential. This sense of inferiority in innovative R&D activity has persuaded Japanese government and industry of the strong need for promoting government-university-industry cooperation.

OPPORTUNITIES FOR COLLABORATION

It is a very good idea theoretically to cooperate in industrial research, but it is very difficult practically in a highly competitive industry such as the electronic industry. But strong pressure from public opinion in favor of better cooperation in government-supported R&D projects forced the government and industry to find a solution. The idea was that it may be possible to cooperate if the area of cooperation is limited within a very basic portion of frontier technology. There are many possible alternate technologies to study and evaluate, and from which to select candidates for future product development. Even after the basic technologies are roughly established, enormous development efforts are still necessary to bring forth commercial products. There is ample room left to compete with one another. So, the VLSI Cooperative Research Laboratory limited its activites to a basic segment of frontier VLSI technology, and technology closely related to the specialized know-how of a company was left as that company's responsibility. In the early days of laboratory operations, research members from competing companies had mixed feelings and communication was very difficult. But after the research projects were clearly established, the management philosophy influenced the laboratory members as well as the managements of member companies; communication barriers rapidly disappeared and very fruitful cooperation began.

International cooperation may not be as easy as this particular case suggests, but we have gained valuable experience and the courage to attempt more difficult experiments. Cooperative development closer to the final product in vital industries such as the electronics industry should proceed between private companies purely from a business point of view. Artificial restrictions and regulations in such areas may harm, rather than enhance, the effectiveness and creativity of the R&D process and thus of industrial productivity. However, there are ample opportunities for joint research or exchange of people in areas where knowledge and technology are still immature and where broader exploitation is needed to meet very important social needs.

As developed countries move into the information-intensive society, "C&C" technology – the marriage of computers and communications aided by microelectronics as a middleman – may have a great and beneficial impact upon society. A special feature of this new concept is the way in which research, development, and application of "C&C" technology have sought, whenever possible, to satisfy the needs of human beings and of society, and thus to make both individual and social life fuller and more convenient. "C&C" technology must exist for the sake of humanity, and must truly contribute to society. In order to achieve this objective, research and development activity should not be confined to a specific engineering community or to any one nation. There are many new devices and hardware systems to be developed, but the most urgent need is for the accelerated advance of software and systems product engineering.

Software production is increasing at a great pace. Since this activity is still limited to specially trained professionals, there is much concern that the imbalance between the demand and the supply of software workers may limit progress in microelectronics and hence in "C&C." Internationally standardized software engineering technology and production tools are badly needed. The development of translation machines would help minimize international language barriers. Accessing and exchanging scientific data and information more effectively is important for an integrated service digital network. For this, the international standardization of protocol is essential.

The future of electronics is very bright, but it is very difficult, if not impossible, to provide the human touches. Our modern society has been losing human touches that are imperative to enrich society. Technology in the past has had a tendency to sacrifice such human touches in order to provide more goods and services at a reasonable cost. We see now a bright chance to change this trend by the advancement of microelectronics, which may relieve manpower for humanistic services. This aim is very difficult to accomplish and can only be done by our sincere desires and efforts. In this respect, there are numerous opportunities to cooperate, and we should look into every means of cooperation for the betterment of our society.

13

Overview of Policy Issues: Panel Report

E.F. deHaan
Richard N. Langlois

Electronics is the glamour industry of the coming decade. As such, the telltale signs of popular mystique are already beginning to appear.

On the one hand, this should make us wary in approaching questions of business or government policy in electronics – questions such as the desirability and appropriate nature of international agreements for technical cooperation. But, on the other hand, we should not ignore the truth behind the mystique: electronics is in many ways unique among today's industries.

One manifestation of this uniqueness is that, unlike the situation in many other sectors, there is a decided lack of cooperation among nations on electronics research and development.

In view of the strong competitiveness of electronics today, this lack of cooperation may be both desirable and inevitable. But there are also indications that some kinds of R&D cooperation among firms can prove beneficial, as in the Japanese VLSI experience so well described by Dr. Uenohara in his chapter. It seems most sensible to think that both competition and cooperation have a place. And our problem lies not in choosing one or the other, but in knowing precisely when and how to add cooperative elements to the competitive process.

KNOWLEDGE AND MONEY

A good starting point is to recognize that the process of competition and cooperation have something in common: they are both about knowledge and money.

We often tend to think only of the money side of this process. In the case of competition, for example, we see what is basically a motivational system – a striving after money whose virtue lies in some superior ability to elicit productivity and hard work. But a competitive process is also a striving after new knowledge; more to the point, it is a decentralized search for knowledge that often has a centrifugal tendency to diversify not only the R&D approaches taken, but also the knowledge ultimately gained.

It is perhaps not surprising that the panel participants found a computer metaphor to express this idea. Rather than conceiving the future of electronics as a single "program," its huge potential might be most fully achieved by a decentralized "distributed processing" approach.

Similarly, cooperation is also about both knowledge and money. When we think of cooperative efforts in industry, the notion of economies of scale often comes to mind. Some activities are cheaper if (and often impossible unless) they are carried out at a scale that requires the pooled resources of two or more firms or countries. But simultaneously, cooperation involves a sharing of knowledge, a pooling of ideas as well as money.

Needless to say, pooling ideas can have its benefits. In some cases, cooperative research and development can produce a kind of "cross-fertilization" effect that improves the individual creative potential of the participants. This effect is particularly important in areas of fundamental research and so-called generic technology, and it is part of what underlies our culturally ingrained belief that cooperation (and, indeed, government support) is the appropriate mode in which to undertake basic scientific research.

This cross-fertilization effect seems also to be at work in Dr. Uenohara's portrayal of Japanese cooperative VLSI effort. Although there were surely some economies of scale involved, the motivation for the multi-firm project, in Dr. Uenohara's view, lay in a desire to replicate, in a country with a tradition of low worker mobility, the kind of technology transfer and knowledge spreading that other countries experience through more frequent job changes among engineers. The fact that this Japanese VLSI project involved a relatively small monetary cost (what some panel members termed "peanuts") tends to confirm Dr. Uenohara's interpretation.

This is not to say that all or, in fact, any cooperative research efforts have an effect that is solely in the nature of cross-fertilization. The overriding tendency of knowledge pooling is centripetal – it brings the ideas and information of the participants closer together. As such, cooperative efforts may sometimes have a distressing side effect: a reduction in the diversity of ideas and approaches necessary for the productive exploitation of technological potential. This problem would likely prove most serious nearer the applications end of the R&D pipeline, where specific products are developed for market, rather than in the neighborhood of fundamental science or generic technology.

ELECTRONICS AS FRONTIER

In considering the role of international cooperation in R&D, then, we are faced with a number of delicate trade-offs.

Resolving the matter seems easiest in areas of basic research and seminal technology. Cooperation in this area allows us to take advantage of some very real economies of scale without seriously reducing diversity; for, although such cooperative efforts narrow the portfolio of <u>research</u> approaches, a successful collaboration can yield a multiplicity of eventual <u>development</u> approaches.

As we move into the application area, though, the problem becomes more difficult. Here we find both greater economies of scale to be gained — in the mounting of expensive development projects — and, at the same time, a greater potential benefit from a "multiple-paths" approach. The question is, which effect predominates?

The answer has to do with the issue we touched on above — the uniqueness of modern electronics.

Although there are economies of scale in many phases of electronics R&D, the characteristic of overriding importance in the industry is its dynamism, its newness, and its expanding technical and marketing boundaries. This argues for a strong emphasis on decentralization in electronics and a deemphasis of competitive R&D whenever the frontier is unknown.

This contrasts, perhaps, with the situation in an industry like energy, in which the promising technologies not only require large capital commitments for R&D, but also fall into fairly well defined technological and marketing categories. (Even so, the benefits of a decentralized approach to R&D in energy may often be underemphasized.)

Of course, there is an argument that although the electronics industry has been characterized in the past by a relative absence of important economies of scale, this situation is now changing. Now that chip technology is becoming better known and more clearly defined, the competitive focus is on cost reduction through mass production; making chips is now a capital-intensive business whose success depends on volume. Does this not suggest the desirability of cooperative efforts to develop chip-making facilities?

The likely answer is: not necessarily. While the hardware aspects of the electronics industry are certainly becoming more capital intensive, they nonetheless represent only a small part of the cost of electronics products. The real area of concern for the future can be summed up in one word: software.

The name of the game is increasingly "the system" rather than "the hardware." As such, there will be an increasing rather than a decreasing number of market niches, with expanding opportunities for an international division of labor. Furthermore, since computers are increasingly used to help generate software, chips can be though of as an input to its production. This means that a software-intensive international electronics industry is best served by the low-cost, high-volume production of chips by those most able to do so. A single facility in the United States might soon be adequate to produce all the chips that country requires; and a single 747 — to use a more vivid image — might carry to Europe all the chips that continent could use for the next few decades.

The implication is that countries feeling themselves behind in the development of chip production facilities should not look to cooperative R&D to "catch up," but should turn their attention to the production of software using the benefits, as one panel member put it, of "chip labor." The crucial strategic question for Europe, which feels itself lagging Japan and the United States in production technology, is this: If we have lost the hardware battle, how do we win the software war?

This raises the issue of cooperative efforts in software development. In fact, there does seem to be a concrete trend in that direction. But, to the extent that software development is an area of burgeoning possibilities with few important capital requirements (that it is "an art, not an industry," as one participant phrased it), the benefits of decentralization would seem to predominate.

One exception is the area of large-scale "system" software, such as that involving two or more countries for, say, telecommunications. Here cooperation is obviously essential. But the kind of cooperation needed is less that relating to R&D than it is cooperation on standards and protocols.

To put it another way, there is a need to standardize the connections among users and countries, leaving the major part of the software as a "black box" that can develop in diverse, competitive ways.

THE ROLE OF COOPERATIVE AGREEMENTS

In summary, then, we can point to two areas in which international cooperative agreements might be of the most value: in basic research and in points of standardization.

Basic research. In areas of basic research, or occasionally of important technology, the usefulness of international cooperative agreements would seem most clear. One often hears it suggested that Europe is lagging in good, basic technology. To the extent this is true, the cooperation of a European concern with a Japanese or U.S. firm could be of great benefit.

Points of standardization. International cooperation should also prove valuable in aiding the development of software through agreements on standards and protocols. Such standards should enable software systems to interconnect and work together without constraining – by overly prescriptive or too early standardization – the potentialities for innovation.

PROBLEMS AND CONSTRAINTS

The possibility of such joint agreements will depend on a number of factors, which differ from country to country.

In the United States, the perception is that joint ventures among companies (across national boundaries or otherwise) is made difficult by U.S. antitrust policy. In the area of basic research, though, this problem is largely mitigated. Joint ventures in research do take place with probably the same ease in the United States as elsewhere — except that they take place under the umbrella of universities.

In Europe, the problem may be somewhat different. Rather than constraints against cooperation, Europeans often feel pressure for more collective action than may be advisable. European governments often see technologies with political eyes, as may be the case with large-scale semiconductor process technologies, and seek to bring together their national companies in pursuit of such technologies for a <u>raison d'état</u> even when the economics of comparative advantage are not favorable.

One problem that seems prevalent in all countries is the difficulty of managing multilateral research agreements. Such agreements are often initiated with motives that are not strictly economic or scientific. But even when formed under the best of intentions, such multilateral groups often suffer from various "free-rider" problems: for example, partners without large development resources of their own frequently want the cooperating group to push on into project development, while those partners with strong development capabilities want to keep the project in the basic research arena. Perhaps more importantly, though, such joint research arrangements, when formed through government arrangements, are extremely hard to abolish and often outlast their usefulness.

For these reasons, the panel tended to see as more workable two alternate forms of international cooperation. One is the simple bipartite arrangement, as between two companies. This keeps the political and management problems to a minimum and permits each party to fine tune the benefits of cooperation to a particular interest.

Another useful, and often overlooked, form of international cooperation in research and development is that which takes place within the boundaries of a single entity — the multinational corporation. A firm with research labs and branches in several countries provides a way of transferring ideas and technologies across borders without some of the difficulties of extramural cooperation and transfer. A useful form of international cooperation among governments could be the recognition of advantages to the countries involved from such internal technology transfer within a multinational company and the development of agreements and mechanisms to support such transfer to derive all benefits possible for the participating countries.

III
Strengthening the Technological Base

14

Energy Conversion and Conservation

Rudolf W. Meier

International technical cooperation among the industrialized countries in the field of energy plays perhaps a more prominent role than in any other field of science and technology. This role is determined by the convergence of three circumstances: the economic and strategic considerations related to energy supply; the substantial financial commitments related to energy research and technology; and the breadth of technical resources required in energy research, development, and demonstration.

The world is not stationary, and there are several conditions which will increase world energy demand over the decades to come.

1. World population will grow. Its present growth follows an exponential law with a doubling time of 35 to 40 years. With more people on our planet, more energy will be required.

2. The world average energy consumption per capita is two kilowatts, but three-fourths of the world population must live with less than this value. Aiming at a conflict-free world solution of the North-South problem, as a consequence, leads to another increase of energy consumption.

3. The rate of economic growth and energy consumption among developing countries will accelerate while industrialized countries will maintain more modest growth rates. The degree of concomitant increases in energy demand is much disputed. Predictions by the World Energy Conference indicate a pattern according to figure 14.1, showing the development of energy consumption over forty years to come, for the whole world, as well as for three groups of developing, socialist, and industrialized (OECD) countries.

This projected energy demand cannot be satisfied if we are not undertaking a substantial effort to develop new and improved conversion and conservation technologies (see figure 14.2). Various technologies with large differences in complexity are involved; some have reached a high degree of maturity, with not much margin for improvement, and some are still in the early stages of research. Let us concentrate for a moment on categorizing various energy carriers and the proportion for world energy supply that they will contribute over the next decades. Figure 14.3 represents the obvious, that oil is the dominant source of energy today and will maintain this role for some twenty years. Its marked decline is expected after the year 2000 due to increasing natural scarcity, if political troubles do not precipitate a much earlier decline. Natural gas is a smaller but important contributor following a similar pattern to oil, but somewhat displaced in time. Coal, the second largest primary energy source is expected to decline in its present classical use. New technologies, however, are being developed to convert it into more useful and publicly acceptable gaseous or liquid forms. Through such processes, the coal share is likely to increase. Nuclear energy is technologically ready for large-scale use, but its acceptability in many countries is questionable.

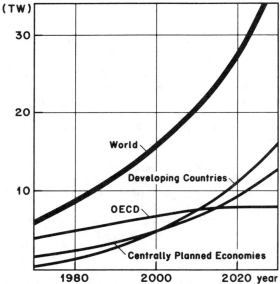

Fig. 14.1. Total energy demand projection.
Source: World Energy Conference 1977

With the introduction of breeder reactors which might be technically and economically feasible by the turn of the century, no fuel supply problem would exist for many generations. Hydropower is expected to increase slightly, but its contribution is limited by resources. The harnessing of solar energy will remain on a modest scale over the next decades, in spite of its unlimited availability. Due to the low power density at the earth's surface, the high plant investments for solar result in energy costs which are uneconomical, within the forseeable future.

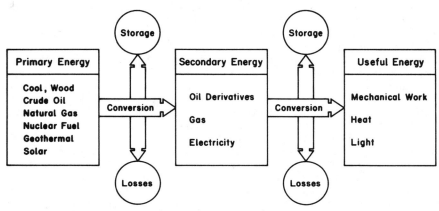

Fig. 14.2. Energy conversion and conservation.

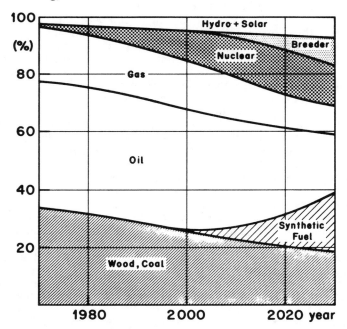

Fig. 14.3. Fraction of different energy sources as part of total energy requirements.
Source: World Energy Conference 1977

It might be worthwhile to point out here that one of the very specific features of energy technologies in contrast to, say, electronics, is its innovation time scale. First, development and demonstration of new energy technologies into commercial operation are both long and slow processes because of the size, costs, and complexity of the systems involved. But still a third factor affects the time scale – the influence of society through political action. Concerns for environmental quality and public health and safety service to filter the introduction of new technologies into society.

Going back to figure 14.3, we might try to deduce arguments as to where priorities should be set for particular energy technologies. Since nuclear and new uses of coal (synthetic fuel) appear to be the highest contributors to energy supply, they are the candidates for the largest R&D effort. This corresponds in fact to how most countries have reacted; yet one has to accept that the efforts should be more diversified to reduce vulnerability to unpredictable national or international circumstances. We have to keep in mind, however, that the picture shown is a representation of the global situation. Locally, for different areas and countries, the proportions look vastly different.

It might be of interest to see how individual countries have allocated R&D money to different energy technologies. Figure 14.4 depicts the relative shares of R&D budgets for Austria, Great Britain, Switzerland, the United States and Germany and the average of the member countries of the International Energy Agency. Only government funding is included. The amount of money funded privately by industry is not precisely known. The largest portion concerns the improvement of classical conversion technologies and would fall under the conservation category. In all countries except Austria, the biggest fraction is nuclear energy, but some years ago it used to be even larger (in relative terms). Coal and new sources or sectors of energy, mostly characterized by "alternative energies," are the next largest. Substantial funds are allocated to nuclear fusion research, even though this technology is not expected to influence the global energy pattern for the next forty years (figure 14.3). Since its introduction, if successfully developed, would meet all energy requirements of the future, a strong effort to demonstrate its feasibility is well justified.

In closing this review of energy conversion and conservation technologies, some of its characteristic aspects can be summarized as follows:

Progress in energy technology is the result of many contributions in many laboratories of many countries.

Energy technology projects require substantial financial commitments because they need a valid demonstration at near full size of a commercial unit.

A consequence of the dimensions of such projects is the long time scale from inception to an operation stage. This is typically one decade or more.

The intense publicity on nuclear energy safety, on air pollution from coal and oil-burning power stations, and the drastic energy price increase by pressure from the OPEC countries all have contributed to create a strong awareness of the public on energy issues. As a result,

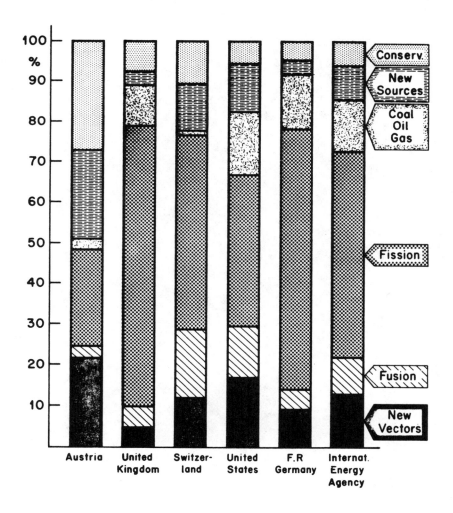

Fig. 14.4. Comparison of government funded energy research in 6 geographical areas.
Source: International Energy Agency, Paris, 1979

the development and introduction of any new energy technology will be a subject of political debate. The persistence of such processes is unpredictable and therefore can affect the time scale of the technical development.

International collaboration in energy technology projects is therefore not an option, but a necessity in order to share costs, personnel, and equipment, particularly in the phase of near full-size demonstration.

15
Overview of Policy Issues: Panel Report
Lionel Boulet

Nations depend on energy not only for industrial development, but also to an increasing degree for agricultural development. As a country industrializes, it constantly increases its demand for energy. If domestic supplies have not been adequately developed, the country must rely on imported energy or reduce its rate of development, increasing unemployment. As the price of imported energy rises, the importing nations must decrease other imports, raise their exports, or devalue their currency and increase the domestic rate of inflation. If we consider that the actual energy supply will be exhausted in less than one hundred decades and that the problem is worldwide, it seems that this "crisis" (as Edward Teller has said) is composed of the symbols for danger and opportunity. Depending on a "true or not" international collaboration, we will move into the future with either opportunity or danger.

Concern for energy supply and utilization permeates the policy planning of all countries. Major efforts are now underway which seek technical advances in energy conversion and conservation. It is not clear what role international cooperation can and should play in all of these areas. The market, and this refers to private enterprise, should be able to bring about more energy efficient appliances and machinery. As the price of energy rises, industries are propelled into a search for more efficient processes and for alternate fuels. Particularly in these last two areas, which most feel have become the heart of conservation and conversion, governments have entered accords which seek to strengthen the underlying research base. Unfortunately, the private sector is not totally present.

INTERNATIONAL MECHANISMS

A number of international mechanisms exist which either undertake or coordinate research in conservation and conversion. Most notably, the

International Energy Agency (IEA) acts as a coordinating arm in assembling interest in various energy projects. The IEA currently derives participation from the United States, the Federal Republic of Germany, Australia, Belgium, Canada, Italy, New Zealand, Spain, the United Kingdom, Switzerland, Sweden, Denmark, Ireland, Mexico, the Netherlands, and Japan. These agreements can provide for exchanges of researchers, the establishment of a direct communications network between researchers, and funds for research. Projects are now underway in such areas as high-energy physics, large-scale wind-energy conversion systems, development of superconducting magnets for fusion power, small solar power systems, the development of geothermal resources, the establishment of a biomass conversion technical information service, and efficient use of energy in cement manufacture. Private sector participation may vary from project to project and country to country. In the United States, for example, industry often participates in IEA projects as a supplier, consultant, or contractor for research. Two United States solar manufacturers are suppliers for a project to design, construct, and operate small solar power systems in Spain. In an international R&D program on superconducting magnets for fusion power, six superconducting magnets will be fabricated and tested at the Large Coil Test Facility at the Oak Ridge National Laboratory. Three will be designed and built by United States industry, and one each will be contributed by Switzerland, Japan, and EURATOM.

Agreements set forth by individual countries on either a bilateral or multilateral basis comprise another major category of cooperation. The United States is a participant in many agreements with a number of partners, including:

- Canada, in the research and development of tar sands and heavy oil extraction
- The Netherlands, in the development of magnetohydrodynamic electrical power generation
- Japan, in the field of high-energy physics and other energy-related fields such as nuclear fusion
- The United Kingdom, in a survey of all aspects of coal use.

The technical scope and mechanisms of cooperation vary among projects. For instance, in the U.S.-Canadian project for R&D of tar sands and heavy oil extraction, the technical areas of cooperation include: (1) parametric analysis of tar sands and heavy oil characteristics; (2) comparative studies of the economics and sweep efficiencies of steam flooding and in-situ-combustion methods; (3) permeability-enhancement methods; (4) treatment and use of water produced during tar sands and heavy oil development; (5) steam flooding with additives; and (6) shaft and tunneling design. The mechanisms include the conduct of joint projects and programs, joint funding of cooperative projects, scientific meetings to discuss progress, and exchange of project and experimental plans for review and comment.

COOPERATION AND ENERGY

If we review the characteristics of cooperation, we must say that the best example was set long ago by research scientists. Hundreds of years ago, scientists, proud of their findings, disseminated their knowledge all over the world, applying the principle "to know and let the others know." With the development of the national industrial society, it became more and more difficult to apply the same principle. The national industrial society believes that to survive, it has to keep all its findings in order to make money on the international market. Not a single industrialized country has realized that there are enough tasks inside to keep all the people employed in trying to adjust outside development for the needs of its own group. We have reached the point where research facilities which are set up in one country and which could serve the needs of many are being duplicated for political reasons.

From the panel deliberations, our group found that there are many areas of common interest where collaboration could bring solutions to many worldwide problems. In fact, the characteristics of energy research worldwide bear on the following observations:

- Energy technology is increasing because of the proliferation of laboratories all over the world.
- Energy projects are large and expensive because the costs of demonstration units are expensive.
- There are long delays before introducing new technology because of lack of collaboration.
- Industrial collaboration, which was an option in the past, becomes a must in the future.

One could divide industrial collaboration in three parts: actual resources, short-term options, and long-term solutions. The actual resources are oil, gas, and coal. Worldwide capabilities should be harnessed in concert to find them, exploit them, and make them available. Maybe the best sources are in the underdeveloped countries, and industrialized countries can help to make them available. In the short-term, until the year 2020, it seems that nuclear energy is the only answer. Many problems are associated with this option which may preclude its use as a transition fuel. To prepare the long-term future, it will be necessary to have international collaboration mechanisms with the help of private industries in the development and utilization of biomass, fission, solar, synthetic fuel, and fusion.

In summary, a review of representative cooperative projects suggests that private sector interest and participation may be greatest at the demonstration phase of technologies that show promise of commercial development. The advantages that bilateral or multilateral cooperation afford are financial leverage and avoidance of RD&D duplication. This appears to be particularly critical in areas of uncertain or longer-term profitability.

Due to the pervasive concern with energy costs and supplies shared among the industrialized countries, energy projects offer special opportunities for cooperation which are of national interest to the countries involved as well as to energy industries. Several models of cooperation exist both for types of participation in financing and research and for mechanisms for private sector involvement.

16
Telecommunications*
Maurice Papo

Because they are on the leading edge of electronic technology, tele-communications systems and devices are becoming part of all the great adventures of the modern world. As a result, telecommunications has become a major element of national industrial policies. In order to understand the potential role for international technical cooperation in the area of telecommunications, we need to understand the trends in national policies as well as the ways in which telecommunications technology itself is likely to affect and be affected by these policies.

Government objectives in telecommunications and telematics are in fact quite similar among industrialized countries. All are concerned with the general development of communications, as it affects both individuals and organized entities. The goal is maximum satisfaction for users at the lowest possible cost. This goal is normally cast in terms of a public-service mission that concentrates on basic offerings such as universal phone service. At the same time, governments are interested in new and innovative services that will capitalize on new technologies and on the convergence between telecommunications and data processing to provide efficient, productive tools for business and industry. Another principal goal is to enhance national competitiveness in international markets by strengthening telecommunications and related industries. In addition, governments often see telecommunications as a route to such ancillary goals as reducing unemployment and providing an alternative to energy-consuming travel.

But while the goals may be the same throughout the industrialized world, the programs and policies implemented by various governments to reach these goals are often quite different. This is most clearly illustrated by two dynamic examples: France and the United States.

FRANCE

The fantastic development of the French telecommunications industry during the past five or six years has been the result of an activist government policy, one stimulated increasingly by the PTT itself.

In the Sixth and, even more so, the Seventh French national plans, the government assigned a priority role to telecommunications. Between 1976 and 1980, some 120 billion francs were invested in telecommunications. As a result, the problem with basic phone service that had long plagued the country has now largely been solved, albeit by what might be called a "brute force" approach. In 1981, the program calls for 26 billion francs to be invested, with 3 billion going to R&D.

The national government, the PTT, and other government entities are also advancing a number of related action programs. These include the December 1978 "Plan d'Informatisation de la Société," and a telematics plan that involves such technologies as videotex, electronic directory services, teletext, the Telecom-1 satellite, fiber optics, and the "wired city." This may also constitute something of a "brute force" approach to technological change. (I should note that, whatever the virtues and disadvantages of this approach, the coining of the terms "informatisation" and "telematique" (telematics) at least underscores the well-known French superiority in taxonomic matters.)

The institutional structure under which these developments are taking place is governed by an 1837 law that has undergone no fundamental changes and may rightly be considered outdated. This legal framework is quite liberal, in the sense that it accords the government broad discretion for involvement in and the management of communications. The current situation is that a government department, the PTT ministry, exercises the telecommunications monopoly. The ministry's relationship with private industry is less one of control than of "tutoring," not only through the ministry's R&D organization (called Centre National d'Études Télécommunications), but also through government plans for restructuring the industry and for government participation in foreign sales efforts.

This ministerial autonomy has held sway for some time. Only recently, and mainly under pressure from the press, has the parliament requested a general debate on telematics. But now, some even see an "orientation law" on the horizon. Recently the government announced the formation of a special commission assigned to monitor the telematics experiments of the PTT.

THE UNITED STATES

The American telecommunications industry has long been dominated, not by a government-owned PTT, but by government-regulated private companies, with AT&T being by far the largest of these. In 1968, the well-known Carterfone decision by the Federal Communications Commission (FCC) – which permitted for the first time the use of terminal equipment not owned and provided by the companies of the regulated phone system – began a process of evolution in the American institutional and regulatory framework.

The main direction of this evolutionary process is toward deregulation. Interestingly, the motive for such deregulation seems to be precisely the motive used to justify regulation in the early days of telegraphy and telephone: the public interest.

Most of the deregulatory efforts have been actions of the FCC itself, but Congress is also turning its attention to deregulation. The result has been a new and unique regulatory environment. Regulation is limited to basic transmission services, permitting free competition among the many telecommunications services that could tap into the basic network. Indeed, the new environment is permitting the introduction of competitive mechanisms not only into deregulated areas, but even into areas that remain regulated.

Congress is now considering, and is likely to pass, some form of amendment to the basic 1937 Communications Act that governs the FCC. Such a bill would seal the new evolutionary pattern.

The relaxed regulatory environment in the United States has led to a new industry – the so-called interconnect industry – made up of hundreds of new industrial and business firms. The FCC has registered some 4,000 products (in the 1976-1979 period) offered by this industry. Furthermore, a new set of carriers has sprung up to offer basic telecommunications services in rivalry with the established utilities. These "specialized carriers" saw their business expand dramatically from less than $1 million in sales in 1973 to more than $263 million in 1979, and that is probably just the beginning.

OTHER EXAMPLES

To a lesser extent, we see the same general trend toward deregulation or liberalization – meaning a relative decrease in the scope and powers of state-controlled monopolies – replicated in several other countries.

In Germany, the Monopolies Commission is investigating the state PTT monopoly, the Deutchesbundpost (DBP). A major concern is the unclear nature of the DBP's cross-subsidization between basic services and the provision of customer-premises equipment like terminals and PBX's. The questions of value-added services and of shared use and resale – major items of past controversy in the United Sates – are also receiving attention.

In the United Kingdom, the recommendations of the three-year-old Post Office Review Committee (also known as the Carter Committee) have been gradually implemented since the installation of the Conservative government. Not only have telecommunications been separated from postal services, but the boundaries of the state monopoly are being pushed back to allow the competitive supply of subscriber apparatus and of value-added services. This reflects the influence of the American FCC's policies.

IMPLICATIONS FOR INTERNATIONAL TECHNICAL COOPERATION

These differences among the policies of the various countries will generate some interesting problems in the international arena.

This is particularly true because telecommunications is becoming increasingly international. There seem to be three reasons for this. First, international traffic is growing faster than national traffic, and this has been so for a number of years. Second, equipment sales in the past were limited to a national level, or at least to the level of the developed world. But new hardware technology, which is drawing telematics closer to data processing, is expanding the business opportunities for truly worldwide sales. International telecommunications standards are also mitigating in this direction.

Third, telecommunications satellites are a new technology whose use cannot be confined to a national territory the size of a European country. Thus, the use of such satellites will become an international matter whether a satellite is sponsored by several countries or by a single one.

Telecommunications are still not included in the General Agreement on Tariffs and Trade (GATT) negotiations. But there are political forces – in the EEC, for instance – that are pushing in this direction.

The implication of growing internationalization is that increasing coordination, even cooperation, will be needed not only between governments and governmental agencies, but also between such agencies and private bodies (or organizations representing private bodies). This is necessary in the interest of effective business communications and in the ultimate interest of the end users themselves.

In Europe, the European Commission on Post and Telecommunication (CEPT) has been engaged in a program of harmonization for several years, although we still see no uniform European telecommunications policy emerging in the near term. More interactive, concerted mechanisms should therefore be put in place to bring the equipment and service industries together (along with the users' associations).

International conferences and volumes like this one may have a role in bringing about such interaction and cooperation by helping to define problems and to explore possible cooperative mechanisms that can aid in their solution.

17

Overview of Policy Issues: Panel Report

Lee L. Davenport
Richard N. Langlois

Those who study the history of technological change tell us that episodes of rapid progress are often the result of "technological convergence" – the revolutionary application to one field, of technologies developed in an entirely different field. This seems to be a good description of the situation today in the field of telecommunications, where the application of modern computer technology is fomenting a visible and much-discussed transformation.

It is well recognized that the changes now characterizing the telecommunications industry are in large part responsible for the relative infrequency of agreements for international technical cooperation in this field, particularly research agreements. But what is less well recognized is that the nature and desirability of cooperative arrangements is being affected not merely by the fact of rapid change itself, but by the direction that change is taking.

In the past, telecommunications networks – a phone system, for example – required costly and specialized equipment. With the increasing use of computers, these characteristics of cost and specialization are changing. The mechanical rotary switching system, for instance, is not an item manufactured in such numbers that notable economies of scale in mass production are possible; furthermore, such switching systems are costly to install. But computer switching relies on a technology whose usefulness goes well beyond the switching function and which is therefore mass-produced at a more efficient scale. And, since the installation cost of computer switching is lower, a larger fraction of the value added in the production of switching equipment now takes place in the factory. All of this means that telecommunications equipment business is shifting from what was predominantly a local market to what is increasingly an international market.

Thus, in telecommunications, the past may not be entirely as relevant to the future as it is in other industries. This is particularly apparent when we consider the changing role of government.

Historically, world telecommunications has been a strongly governmental affair. In most countries, in fact, it is the province of a government monopoly – referred to almost generically as a PTT – that manages not only the national phone system, but very often the postal system and various other communications services. Even in a country like the United States, where phone service is provided by private companies, the industry has been tightly regulated at both the federal and state levels. (As one panel participant put it, AT&T is itself "almost a government.") Under government ownership, the government sets the service and financial goals for the national communications system, it undertakes the purchasing of equipment, and it often conducts research in its own laboratories.

This approach developed in a world in which communications technology was centralized, specialized, and costly. But that world is changing. The equipment the governments are now buying for telecommunications is based upon the same equipment the private sector is buying in large quantities for many other purposes. And, as the potentialities of this technological convergence of computers and telecommunications become more apparent, the uses to which telecommunications systems are put will be determined less by the constraints of technology than by the tastes and budgets of the users. All of this suggests the advent of a regime to which the centralized PTT approach may be less well suited.

There are in fact trends in some countries that seem to confirm this speculation. In the United States, as M. Papo suggested, the trend is toward deregulation, particularly with regard to the user devices that can be connected to the telecommunications networks. On the other hand, the planning and direction of telecommunications services remain, in some countries, under close government control. We can think of the United States as having a sort of post hoc planning system, in which the market calls forth innovations and determines the shape of telecommunications services and the government regulatory mechanism comes into play only after that shape is clear. This contrasts with the more ante hoc approach in Europe, where government attempts to plan the development of telecommunications services and guide the national role. A good example of this might be the current French attempt to develop a low-cost computer terminal – a national strategic gamble in R&D initiated and guided by the PTT.

There was some sentiment at the panel session that these two approaches were not ultimately different in their effect. But which, if either, is a preferable approach must remain an empirical question.

At base, this is a question of the relative desirability of centralization of planning and control versus decentralization. And it is a question central to a consideration of international technical cooperation. (This theme was also brought out at the electronics panel session, whose list of participants overlapped significantly with that of the telecommunications session.)

There are two areas we might examine as possible subjects of technical cooperation in telecommunications: research and standards.

There seemed to be a panel consensus that the standards issue remains the more important, or perhaps more relevant, to international considerations.

To the extent that the telecommunications revolution involves phenomena of technological convergence, the important problems are those of application and development in an atmosphere characterized by market uncertainty far more than by technological uncertainty. This contrasts with a situation – as, perhaps, we may face in the field of energy – in which the product is well defined and the market potential clear, but the technological problems are the source of uncertainty. The case for cooperative research is much stronger in the latter instance than in the former.

Research in telecommunications and in the basic electronics that underlies it is largely a strategic problem – for governments as well as companies. From the national point of view, the question is: Given that the technical potential is so vast that we, as a single country, will never be the leader in all areas, which pieces of the technological puzzle should we address? Countries are concerned with employment arising from the flow of basic knowledge, and this is why the location of corporate research facilities is an important issue. But most countries (especially in Europe) must recognize that a very large percent of all basic knowledge will always originate outside their borders. The strategic problem is slightly different for a private company, but it is no less important. Here the issue is what we might call the "nationality/multinationality" problem. To what extent can a company be multinational and still preserve its national identity? Are national and corporate interests at odds in preserving a distinct "national" company at the expense of the technology-transfer benefits of unrestricted multinationality? Or is a company well served by preserving, say, its "reputation for being French"? The issues, in any case, are strategic ones.

Standards are another matter, and here we see a clear dilemma. Since telecommunication, like all forms of communication, requires some common ground, some agreement, some systemic uniformity in order to function at all, international standards – in both technical protocols and legal or institutional arrangements – are extremely important. Yet, such standardization in a world of rapid technical change can have a stifling effect on innovation if it comes too early.

In the end, standards are what one makes of them. They can be used as trade barriers. They can often be used as ways of exploring common ground in an effort to initiate trade. The International Telecommunications Union (ITU) has long concerned itself with standards. But there may be some reason to think that, because of its relatively political orientation, the ITU may not be the best forum for all discussions on the setting of standards. Indeed, international conferences may have an important role in bringing the various groups together to discuss international standards as well as in helping to sort out the intellectual issues involved.

In sum, standards rather than research should be the important focus for international cooperative agreements in the area of telecommunications. There are ample opportunities for such cooperation. And, while some standardization is essential, standards that are applied unilaterally or that come too early pose threats to the pace of innovation.

Many intellectual issues remain. We need a better idea of what has changed in the environment facing the telecommunications industry. This would give us a better idea as to which parts of the developing international system need to be handled at national levels, which at international levels, and which are best left to the private sector. We also need a better understanding of standards and their effects — of which parts of the system should be standardized and which should be left flexible. At which point in the innovation process, should we begin to set standards?

18
Transportation
Ake Zachirison

STRUCTURE OF THE TRANSPORTATION SECTOR

International cooperation is most effective when there are broad problems common to many nations, and the results of such cooperation can be applied to social and economic needs. Many of the characteristics of the transportation sector surely meet these criteria.

We should differentiate between basic responsibilities of governments for the operation of systems — infrastructure of roads, speed limits, standards, conditions of energy use — and the responsibilities of the private sector for the development of the physical equipment and all of the components, control devices, and energy conversion systems that constitute any mode of transportation (see figure 18.1).

Let us briefly examine the scope of transportation as it has evolved in recent years. The United States spends about 20 percent of its gross national product (GNP) on transport, Western Europe spends about 15 to 16 percent of its GNP, and developing countries spend less than 10 percent.·In a similar way, there is a close correlation between consumption of energy and GNP, and between vehicle registrations and GNP per capita. Moving into the future, the road and rail system in Western Europe will be a limiting factor for more cars and trains. For the next twenty-five years, an average increase of only 35 percent in the length of major roads and construction of only an additional 5,000 kilometers of motorways is predicted for the whole of Western Europe. This implies an increase of approximately 135 percent in the number of cars and 125 percent in the distance traveled over the same period.

NATURE OF THE TECHNICAL BASE

The breadth of technology called upon is enormous. It ranges from the processing of materials in construction, through conversion, and to application of energy to avionics and systems for safety and communication. Transportation is becoming a major consumer of advanced computer technology.

In more detail, progress in transportation rests upon technical progress in a number of critical areas: energy supply, emission control, minerals and material substitution, safety requirements, legal actions from governments, and labor costs.

Cost/Trip versus Passenger/Hour

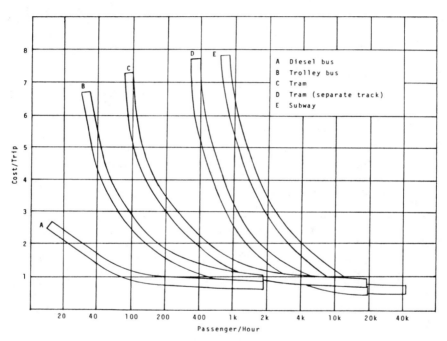

Fig. 18.1 Current vehicular cost/trip ratios.
Source: The Royal Academy of Engineering Sciences, Sweden (IVA),
           "The Motor Vehicle 1980-2000," Report No. 215, 1977.

Clearly, the most critical problem facing this particular industry is that of energy. Allocation of the several key sources of energy to the principal needs of society is one of the most important strategic policy considerations of any government – particularly those of the OECD countries, because they have so many demands for energy use. Those responsible for transportation development must work with certain assumptions of fuel availability, by type and by quantity, and must establish technical programs compatible with these assumptions.

These considerations will form the context for decision making in the foreseeable future. Energy supply, and especially the availability and cost of oil supplies, will determine the particular configuration of policies that impact transportation.

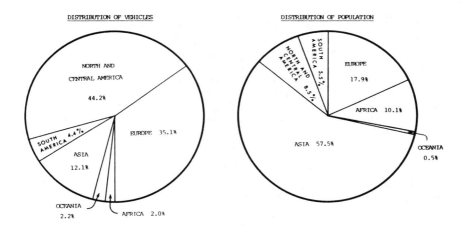

Fig. 18.2. World distribution of motor vehicles.
Source: MVMA (Motor Vehicle Manufacturers Assn.), U.S.A., Facts and Figures 1980.

According to exploration experts, the total world crude oil reserve ranges between 1,350 billion barrels and 2,100 billion barrels. With the assumption of total reserves of about 1,600 billion barrels, we get a dynamic forecast of maximum production around 1985 and then a slow decline in world production from around the year 2000.

Some experts have called the 1980s the decade of natural gas. Natural gas will be transported by pipelines from Algeria to Italy and also from the North Sea to Western Germany (Ruhrgas AG). There will probably be an extension of natural gas pipelines up to Scandinavia. Norway will transport its natural gas from the northern gas fields in the North Sea by pipeline to the area north of Bergen and from the Tromso area along the eastern Norrland coast of northern Sweden down to the European continent within twenty years.

The substitution for oil and natural gas should be another resource of even higher energy density than when oil was substituted for coal.

Uranium is such a resource, which the world needs for nuclear energy during the two critical decades ahead to facilitate the transition from oil to other forms of energy in the next century.

Methanol, synthetic gasoline, and diesel oil are the most probable alternative motor fuels in the near future. For aviation, however, hydrogen will be the most likely substitute for JP-fuel. Methanol may not be the cheapest alternative motor fuel, but together with diesel oil it may become a second alternative to gasoline and thereby minimize the risk of a supply interruption of one fuel.

Another important energy conservation measure should be technical changes on the car itself because of the high cost and long impact time. It will not be possible to introduce alternative engines, like Stirling engines or gas turbines, in large quantities on the European car market before 2000, because the capital needed will be directed to energy solutions. One efficient power for cars is the electric motor, driven by either batteries or fuel cells. The fuel-cell-powered electric motor represents the second class of future car engines suitable for mass production. However, the development of the fuel cell has not reached a state where it is sufficiently competitive in price, weight, volume, or fuel conversion efficiency to be commercially feasible in cars today. Battery powered electric cars will be introduced before 1990, but only in urban areas and for fleet use in limited quantities because of a restricted range coupled with high cost.

PROBABLE TRENDS IN TECHNICAL DEVELOPMENT

Based upon these considerations, one could devise a list of trends and constraints for the transportation sector in the next two decades. This list would include three major concerns – economic growth, energy supply and utilization, and societal requirements. The specifics of the scenario would be as follows:

- Average economic growth increases by 1 to 2 percent during the 1980s.
- Energy consumption growth increases 2 to 3 percent over the period.
- Annual production (consumption) of oil remains at relatively constant levels until the end of the 1980s.
- Diesel oil share of petroleum increases from 30 percent to 40 percent.
- Methanol from natural gas will be available in commercial quantities about 1985. Ethylene for plastics will be derived from natural gas instead of from oil.
- High-temperature pilot reactors will start to produce limited amounts of hydrogen or methanol during the 1990s.
- Average fuel consumption for Western European cars will be 0.70 liters/10 km by the year 2000.
- Safety and emissions-control problems will lag behind the overall energy problem.

- Public transit will grow to 30 percent of the transportation sector.
- Buses will increase in production. Minibuses (van pooling) will be more common, together with conventional sized buses.
- Car pooling will increase.
- Minicars will expand their share of the market from 25 percent to 35 percent.
- The diesel engine passenger car share will grow from about 4 percent to about 10 percent of the market. Spark-assisted diesels and ceramic coating in cylinder heads will be advanced.
- Electronics will be utilized mainly to limit fuel consumption.
- The aerodynamic drag coefficient will be reduced to 0.30 on new cars.
- LPG- and natural-gas-driven taxis and buses will operate in limited fleets.
- Urban battery electric cars and vans will operate in limited fleets.
- Serious development of fuel-cell electric cars will be underway by the end of the century.
- Trains will take an increasing share of passengers from cars.
- Development of pipelines will accelerate primarily the use of gas and also coal slurries.

In brief, the energy situation will have a great influence on transportation. Governments will issue regulations and laws that will affect the choice of transport. Travel modes and transport systems will change. New or more energy-efficient transport means will be developed. Regulations for cargo and passenger transport will be standardized through international mechanisms.

OPPORTUNITIES FOR INTERNATIONAL COOPERATION

The nature of the transportation sector, and the breadth of technical problems described, appear to offer a rich field for constructive international cooperation. The competitive aspects of specific designs and technical approaches can be separated from a number of common interests in materials, energy, safety, and so on. There is, in fact, good reason to seek further interactions between public and private efforts which will aid the objectives of both while respecting the independence needed for competitive advances.

Many activities already exist in the public sector. One noteworthy example of international cooperation is the Experimental Safety Vehicle (ESV) program which was begun in the early 1970s and involved the cooperation of Germany, Japan, France, Italy, the United Kingdom, Sweden, and the United States. Originally, this program focused on systems specifications for safety technology (crashworthiness, accident avoidance, and post-crash factors). In 1973-1974, the program was modified and renamed the Research Safety Vehicle Program. This involved a reexamination of safety technology performance specifications and was broadened to include engineering advances for fuel economy, consumer costs, and practicality.

Within each country, the private sector automobile manufacturers participated in the program under different conditions and for different purposes. About once a year since the early 1970s, an international conference has been held to exchange information on each country's progress. In addition to representatives from the governments involved and the participating automobile manufacturers, attendance at these conferences has included representation from insurance companies, professional societies of automotive engineers, transport research institutes, and international organizations. This interest and participation has served to facilitate the dissemination of useful information and assimilation of engineering advances.

Conferences of this type are of great value to identify the problems and possible solutions within an international context. Other cooperative projects span the transportation field from magnetic levitation research to air traffic control.

COOPERATION: CONDITIONS AND INCENTIVES

Due to the international character of the transportation network, there are many opportunities for the development of intergovernmental programs that might in some cases involve industrial participation.

The international implications of issues affecting transportation are being analyzed through existing international organizations like the Common Market, the International Standard Organization (ISO), the Committee of Common Market Automobile Constructors (CCMC), the Industrial Symposium on Automotive Technology and Automation (ISATA), and others. These organizations are working on certain aspects of the problems, trying to find common solutions that can be accepted by all countries.

However, their cooperation does not cover all the issues. For example, in the automobile industries, most Western countries are preparing laws and regulations for emissions, crash safety, alternative fuels, and so on. Even if there is knowledge to some extent of what others are doing, the work is not coordinated. Thus, there is a risk that different regulations will be issued in different countries. The effect would be that an automobile crossing a border would face standards that it cannot meet, and consequently the car may not be allowed to enter. Since most automobile industries have an international market, it would also mean that cars have to be designed and produced in special versions for each country. The cost of vehicles will consequently be higher because of such specialization. International cooperation can therefore greatly facilitate the resolution of many of these problems.

The Western World automobile industries will invest over $100 billion in R&D and manufacturing plants for the next five to ten years in order to meet the new demands and constraints. The cost involved makes it natural for the industry to seek partners to share the economic burden, and therefore an increasing amount of cooperative projects between two or more companies are being established. Fiat and Peugeot recently made public that they will develop a new small gasoline engine. Six major European automobile companies (British Leyland, Fiat, Peugeot, Renault, Volkswagen, and Volvo) signed an agreement in 1980 which aims at seeking and eventually performing major R&D projects of common interest. Peugeot, Renault, and Volvo have been manufacturing a six-cylinder gasoline engine together for many years.

It is now time to answer more specifically the major question put to this conference: Can international agreements between two or more countries improve industrial R&D? In the Western World, one company doing R&D can, without government involvement, establish contacts with industries in other countries to improve its R&D capacity. Also, if similar R&D work is considered in another country, a cooperation agreement may be made to pool resources and avoid duplication. An example of this is the French-British Concorde project. It seems that the general experience is that agreements between Western countries do not affect the R&D in the transport industry unless they deal with very expensive projects. Highest demand for formal agreements between countries involve cooperation between a Western country on one side and an Eastern or developing country on the other. One major reason for this is that an East European or developing country requires a government agreement in order to establish any contact at all.

In the Western World, international agreements and joint R&D are necessary to solve common problems. They can also decrease cost and increase the reservoir of science and technology. Governments do not need to be involved unless very expensive projects are considered or where the R&D work is the basis for laws and regulations.

19
Overview of Policy Issues: Panel Report
Bernard Schmidt

Transportation and traffic are of great importance for society. Both are the basis not only for the provision of goods and services, but also for communication and mobility. Therefore, the government essentially directs the development of policy in this field. But at the same time, transportation and traffic are a significant factor of industrial production. Therefore, international cooperation in this field is determined by cooperation of the governments as well as cooperation between industrial enterprises. In many countries, due to the public importance of transportation, research and development as well as investments are supported by governmental funds. The objectives are always similar:

- Better service for the user
- Economical operation of facilities
- Decreased total costs for transportation and traffic services
- Improved use of raw materials and development of environmentally acceptable systems to reduce adverse impacts on society and the environment

Government support refers to almost all aspects of transportation: automobile transportation and road traffic; public urban transportation; rail; transportation of goods; maritime transportation; and aviation. This broad topic was studied by the panel in the context of international technical cooperation, and we focused our attention on selected critical areas.

AUTOMOBILES, ROAD TRAFFIC, AND PUBLIC URBAN TRANSPORTATION

Automobile production is a mass production. The producers sell their products in highly competitive markets. This is the reason why the penetration of new technologies occurs very slowly and does so only when it supports the competition of enterprises or is required by governmental regulations (such as emissions controls). International technical cooperation is only possible where the interests of industry are represented. Governmental cooperation exists primarily in areas where the standardization of regulations is concerned, such as in determination and evaluation of emission standards.

Public urban transportation is always subject to cost pressure. Furthermore, its low attractiveness has fostered user preference for individual transport. For this reason, technical development is supported by governmental funds in many countries. For example, in the United States, France, and Germany, new kinds of monorail systems have been developed. However, because of the novelty of such systems and their relatively high costs, resistance against the introduction of these systems to the user is very high.

Increasing significance must be attached to the improvement of bus systems. Similar-development works are being carried out by different countries, such as dial-a-bus systems, laterally guided buses, or dual-mode buses. Here, possible cooperation may also be initiated only by industry itself and will be guided by necessities and possibilities of cooperation in the conventional bus sector.

The cooperation of the transportation authorities can also be of significance in this sector for the future. A standardization of requirements may facilitate the introduction of new technologies. First efforts are being made here within the framework of the European program, Cooperation in the Field of Scientific and Technical Research (COST).

CONVENTIONAL RAIL

Rail, up to thirty years ago, was one of the supports of economic progress. However, it has lost great portions of transportation revenues to automobiles and air transportation in almost all industrial nations. For example, in Germany, the proportion of rail usage in the transportation sector decreased from 38 percent to 7 percent between 1950 and 1970. This is true for passenger transport as well as for transport of goods. Due to today's energy situation, specifically with respect to the oil market, and due to concerns for environmental quality, many transport specialists think it is necessary to increase the proportionate share of railway transportation. The precondition is a competitive offer, which is only achievable by technical progress and modernization of the mobile equipment and infrastructure as well as by organizational and logistic improvements.

Due to their unfavorable economic situation, many railroad authorities can make such improvements only to a very limited extent. Research and development programs have thus been started in most industrial countries, with the following objectives: higher attractiveness than individual transport, due to traveling speeds up to 200 kilometers per hour; lower operation costs for goods transportation by reduction of "wear and tear" on railroad and car, achievable by technological means; and improvement of the integration of railroad goods transportation in the whole transport system, resulting in higher transport speed or reliability. This also includes new technological developments, such as lifting devices and transportation machines for transfer procedures and engines with dual-mode locomotives.

Numerous test and demonstration plants have been built and operated for development and test reasons. Some notable examples are:

- The railroad test plant in the United States (Pueblo), where endurance tests are carried out with overloaded goods trains
- The test plant of British Rail in Derby
- The test plant Scherbinka near Moscow, where endurance tests on railroad elements and running tests are carried out, as well as special tests of the behavior of rail steel at extremely low temperatures
- In France and Japan, where new specially designed railroad tracks are used for testing newly developed high-speed trains
- In Germany, where the problems of running technique have been evaluated using complex mathematical procedures.

Verification of these results then takes place on a stationary test rig ("rolling test-rig"), where test speeds up to 500 kilometers per hour are attained. Finally, operation tests are carried out, with test speeds of up to 350 kilometers per hour.

After these tests are completed, analyses begin which permit the design and layout of systems where both technical and economic aspects are optimized. And into the future, to improve the logistics and procedures in goods transport, new installations for the transfer of goods (from one transport system to the next) will have to be designed and constructed. Besides these future-oriented works, there is a continuing process of adapting railroad technology to technical progress, for example, in the field of construction and machine engineering, information sciences, and power electronics.

Almost all programs are controlled by the national railroad authorities, financed by governments or railroad authorities and executed by industrial enterprises and institutes on a national level. Existing international railroad organizations do not have much influence. A prime example of the difficulties involved in joint development and ensuing application is provided by the development history of the automatic clutch, which the UIC has been working on in Europe since 1922. Today, 1995 is mentioned as the probable date for its introduction.

As international traffic attains increasing significance, especially in Europe, a close coordination of the development work with the aim of a possible standardization of systems would be economically desirable. An example of insufficient international coordination is provided by the varying electric power systems, which were introduced into Europe during electrification.

One must also consider the factors that complicate international coordination. Railroad authorities generally prefer to cooperate with national industry, which in the past has developed the railroad technology for that particular country. More or less fixed marketing shares make it somehow uninteresting for industrial enterprises to cooperate with foreign partners for the national market. In the past, there have also been military reasons for autarky endeavors in the railroad sector.

For the market within the industrial nations, cooperation in development and manufacture between governments and industrial enterprises is generally restricted to a limited exchange of information and joint planning investigations in areas where railroad traffic extends across frontiers. For the market of Third World countries, on the other hand, the stiff competition, a certain overcapacity, and the necessity for standardization have already led to successful cooperation between industrial enterprises.

As a final note on this topic, one particularly desirable way to reduce development costs would be better joint usage of test plants in the individual countries, with their complementing possibilities.

HIGH-SPEED MAGNETIC LEVITATION SYSTEM

The development of the high-speed magnetic levitation system is being worked on in many countries, including Germany, France, Great Britain, Canada, Japan, Russia, and the United States. The aims vary. In Great Britain and to some extent in the USSR, applications are seen for local transport, while the development goal in all other countries is a high-speed system for long-distance transports. In Japan and Canada, the research and development is generally concentrated on electrodynamic levitation systems, while in the other countries, electromagnetic levitation systems are of prime importance. The greatest progress here is in Germany and Japan.

The development of such a revolutionary transport system is particularly costly and risky. A direct commercial benefit is not to be expected in a short term for the developing industry. Therefore, the development is funded by the government or governmental authorities. Because of the publicly financed research programs in all these countries, there are favorable preconditions for international cooperation in this area.

The cooperation is necessary for general reasons. A magnetic levitation system as a long-distance transportation system and high speeds of more than 400 kilometers per hour require large operation spaces for economic reasons and make sense primarily for international transport. The introduction of such a system in standardized form in two or more neighboring countries generally demands a prior joint development. Many individual systems with different technical solutions have to be developed. Cooperation contributes to the limiting of development costs in the individual countries.

At present there are already a number of bilateral international agreements – for example, between Germany and France, Germany and the United States, Canada and the United States, and Japan and the United States – as well as continuous joint meetings.

The range of cooperation extends from the exchange of information to the completion of joint development projects. In many cases the direct cooperation of research divisions of industry within the framework of governmental umbrella agreements has worked.

AVIATION

In the field of aviation, the pressures for international cooperation and division of labor are greatest. This is due to the high costs for development and manufacture of aircraft and infrastructural products, the high technical standard which must be attained, the complexity of the technical systems, the restricted national markets and resources, and the international nature of air transportation. Therefore, the greatest number and the most marked examples of international cooperation are found in this field. This is supported by the international cooperation which exists in the military sector within the framework of NATO. Such contacts developed in the military sector also prove useful in the field of civil aviation. For example, the national research tasks and governmental programs are available to the relevant bodies of the European community and NATO. Within the Advisory Group for Aerospace Research and Development (AGARD), an intensified know-how exchange is practiced.

On the European level, coordination regarding national research programs exists within the Group for Aeronautical Research and Technology in Europe (GARTEUR), consisting of government representatives from Germany, Great Britain, France and the Netherlands. Furthermore, jointly initiated research work in selected sectors is controlled by this group. Several years ago, the same governments signed an agreement on the joint usage of existing and planned aerodynamical test plants. One example for this cooperation is the development of a concept for a large European transonic-wind tunnel.

In addition to such multilateral cooperation, bilateral cooperation agreements also exist. In addition to these arrangements, the aviation industry itself is eager to develop stronger international cooperation. The airbus project is a convincing example.

With respect to landing aids, international cooperation is indispensable. With increasing air transportation demands, the limits of the instrument landing system (ILS), which is employed on a worldwide basis, are becoming more and more apparent. For almost fifteen years, the International Civic Aviation Organization (ICAO) has been concerned with the definition of operational requirements for a subsequent system. In 1972, an international contest was started. The resulting recommendations are the basis for international cooperation on governmental research, and industry levels.

20
Environmental Control
W.H. Gauvin

My intention in this chapter is not to provide an exhaustive review, but rather to focus attention on some items of interest by way of illustration in order to stimulate further thought and comments. I would like to caution that my perception of this topic is highly colored by the fact that I am a Canadian, and that the development of new technologies in Canada which are conducive eventually to international agreements on technical cooperation is, in many respects, unique. Industrial R&D in Canada is by and large weak, while the bulk of our national research effort is carried out by very large government (both federal and provincial) laboratories. As director of the Noranda Research Centre, my views are further biased by the fact that we are acting as a profit center for a large Canadian mining company (Noranda Mines Limited), and that our activities are governed by the concepts of risks, profitability, and the far-reaching consequences of the introduction of new technologies (ecological, environmental, social, in addition to all aspects of health and safety in working conditions).

MOTIVATIONS

One important conclusion which I have drawn from the presentations of other conference participants is that each international agreement for technical cooperation is truly unique in character. Nevertheless, one may venture to say that the major driving force behind the objectives of a given international agreement is motivation and benefits. Although this statement may constitute an oversimplification, it would appear that these two factors give each agreement its unique characteristics.

From the point of view of the private sector, the characteristics of an international agreement should be the following:

- It should cover a project of major importance, with identifiable scientific uncertainty.
- It should be non-market-oriented.
- It should be of long-range importance.
- It should probably involve a considerable financial risk.
- The pooling of scientific and technical resources between the potential partners should constitute a definite advantage.

In addition, strong points of motivation for the private sector to enter into an international agreement involving foreign governments would be to facilitate the eventual approval or licensing by these governments of the new products, processes, or other actions resulting from the successful completion of the project. The private sector might also anticipate that such agreements would remove impediments in commercialization and nontariff barriers, and, generally speaking, would facilitate market penetration.

It was also strongly emphasized that the strengths of the private sector reside in its planning and managerial skill and in its ability to conduct realistic techno-economic evaluations. These strengths should in no way be diluted either in the planning or during the implementation of the project.

From the point of view of governments, the major motivating factors are:

- The project offers advantageous political and prestige implications.
- The project may lead to an expansion of trade.
- The project may provide solutions to thorny problems of a societal nature.
- The project may have attractive cultural aspects.

The participants also noted a number of difficulties which may arise during the implementation of international agreements of cooperation. Chief among these were the problems of technology ownership; delays in the implementation of the work, owing to ponderous methods of management, particularly in large projects; and lack of adequate capital investments upon successful completion of a project may be slowed by the slackening of political will, by conflict with activist groups, or simply by a change in the government's priorities.

Turning now to the topic of our panel session, first of all, a consensus is apparent that environmental controls are a painful subject to research establishments whose performance is judged on the basis of the productive innovations they bring into being, and certainly not on the number of ecological or environmental fires they help to put out. And yet the latter activities are absolutely necessary for the survival of the productive operations of the firm. Plainly put, it is a necessary cost of doing and staying in business. That this cost is high is exemplified by the fact that 21 percent of my R&D budget is devoted to the study of ecological and environmental problems, to the improvement of working conditions, to biological surveys, and to the problems of waste disposal.

Neglect of these considerations has led to the almost complete disappearance of the zinc extraction industry in the United States, which was based on the old retort process with its attendant environmental difficulties. In the same country, a number of large copper smelters were forced to shut down during recent years for similar considerations, with resulting losses to the private sector in excess of one billion dollars. In many fields, new technologies had to be developed or purchased to meet increasingly demanding pollution criteria. It is the latter consideration which impelled us, in the mid-sixties, to develop a new process for the continuous smelting and converting of copper concentrates, which will probably affect the survival of a number of threatened copper smelters in the industrialized world. Of all the fields of scientific activities which might be considered for international cooperation, that of environmental control is probably the most amenable to this type of agreement, since it generally involves little consideration of commercial or market competition and is generally perceived by all the parties and by the public as being beneficial from a societal point of view. A number of such agreements are already in force. Others are currently being negotiated. As examples, the following activities can be mentioned:

- Nuclear waste disposal
- Nuclear safety in general (including decommissioning of old reactor installations)
- Control of emission and diffusion over international boundaries of effluent gases containing sulfur dioxiode and dust particles (acid rain)
- Disposal of waste materials in the sea (part of the "Law of the Sea")
- Meteorological monitoring
- Transportation of hazardous materials by sea or by rail across international boundaries
- Licensing of new drugs
- Eutrification of lakes due to phosphates.

These activities form only a partial list of subjects which may well be fruitful areas for international technical cooperation. They do reflect several key criteria which I regard as central for private sector involvement in technical agreements – minimal market competition, long-range importance, financial leverage, and clear technical and scientific advantages.

MECHANISMS

If we now turn our attention to posible mechanisms that could be feasible for industrial participation on government-to-government agreements, those below may be envisaged.

Exchange of Specialists

Owing to the ever-increasing degree of sophistication and complexity in R&D experimental techniques and instrumentation as well as in computerized methods of analytical interpretation of the data, it is evident that the highest level of expertise will be necessary for the successful prosecution of many of these projects. An exchange of specialists with the required degree of competence in their given areas of expertise would appear to be the approach most conducive to successful results.

Sharing of Equipment and Facilities

Many private laboratories, as well as government research institutes, possess laboratory facilities which are almost unique in their capabilities in a given area of investigation. Duplication of these facilities would be unjustifiably wasteful, and sharing would again be the best approach.

Joint Research Projects

The concept of pooling research resources into <u>actions concertées</u> which draw on the research capabilities of the three sectors of the scientific community (private industry, universities, and government laboratories) is gaining increasing acceptance for the solution of problems of demonstrable economic and/or societal importance to a nation and involving a large degree of scientific uncertainty, particularly when a multidisciplinary or multisectoral approach is required.

Cost Sharing

In all three mechanisms which have been discussed, the sharing of costs will be an important consideration. Substantial savings should be achieved by avoidance of costly duplication of facilities, by charging only the depreciation of an existing facility, and by reductions in the duration of a project.

Needless to say, in all these mechanisms, an overriding consideration will have to be given beforehand to the eventuality of acquisition of proprietary knowledge arising out of the work, which might lead to the filing of patent applications. This process, however, has been worked out satisfactorily among other research partners in joint projects on a case-by-case basis, so numerous precedents do exist for accommodating such an effort.

21
Overview of Policy Issues: Panel Report
Part I
Peder Waern-Bugge

THE CHAIRMAN'S PERSPECTIVE

Milton Friedman has said "there is no such thing as a free lunch." For both industry and government, in the area of environmental regulation which has been characterized by negative attitudes and adversarial relationships, this saying has come to have special meaning as both sides seek a more balanced perspective. It was the pursuit of a more balanced perspective that tied together the considerations articulated by the panel.

Because governmental policy is a reflection of society's concerns, much of the discussion focused on the context for regulatory decision making. To develop environmental policies which incorporate well-considered cost-benefit analyses, it is essential that the economic, technical, and ecological issues be discussed and evaluated by all parties concerned in a more dispassionate manner than has been the case over the last decade. Often in the past, industry has adopted a rigid, negative view to any environmental consideration, and in response, special interest groups have lobbied for extreme restrictions. The process has been further complicated by changes in governing administrators. As a result, the trend has been toward short-term and sometimes shortsighted solutions rather than toward more difficult but necessary long-term remedies. Another problem from recent history is that regulatory decisions have been made under intense pressure and without complete evidence, causing substantial and unnecessary loss of technical and financial resources – as the cases of cadmium and cyclamates illustrate.

One problem in a democracy is that a small but vocal group devoted to what we would call "single-issue" politics can make compromise among national objectives very difficult. Along the same lines, it takes great courage for a democratic government to propose solutions that take several objectives into consideration and then to select a course in the national interest that may offend specific groups. This has been done in France, for example, where the government arrives at a decision and acts "autarchically" to override minority vetoes.

The context of the present time demands a more constructive approach. France and Japan provide good models. A critical first step is for both government and industry to inform the public accurately and understandably on complicated technical issues. The true opportunity costs of enacting a specific policy must be carefully deliberated. One of the major costs to industry in the past has been requirements for changes in technology and procedures involving diversion of capital and other resources. The result has had critical effects on productivity growth now affecting the overall economic performance of the industrialized countries.

One of the principal realities that society must confront, particularly in the area of regulation, is that we cannot attain a zero-risk life. We routinely accept the potential hazards of traffic in our daily lives, whereas we may be so overly concerned about the potential hazards of pharmaceuticals that regulatory bottlenecks may deprive people of their lives which could otherwise have been saved with new but not yet authorized effective drugs.

On the other hand, it is important that industry leaders accept the challenge of present-day life. If American car manufacturers had earlier understood the real impacts of changing customer attitudes, new technology, and more expensive petroleum, the transition to newer generations of small cars would probably have come as less of a shock.

Issues of environmental control within a cooperative international framework also raise questions of balance. The agreements for cooperation must be based on a consensus among the parties and on the mutual interests of those involved.

Part II
Lois S. Peters

The global aspect of environmental science requires that the subject be broached within an international context. The number of international agreements related to environmental control reflects this need. These agreements appropriately cover three broad issues: interdependency problems such as transboundary air pollution and multi-national management of water basins and exploitation of the sea bed for minerals and oil; distributive issues of equitable access to natural resources, including provision for reliable sources of energy at affordable prices under acceptable environmental constraints; and vulnerability concerns having to do with hazardous substance production and disposal and development of standards of acceptable environmental risk. The specific subjects of international agreements include acid rain, trace metal and organic material contamination, siting of nuclear power plants, oil spills, health and safety issues in the mining industry, toxic substance control and disposal, and the global effect of carbon dioxide buildup.

The general focus of the panel's considerations was on the question, "What role, if any, should the private sector play in these agreements?" The primary issue of concern was that many of the international agreements are diffuse and unclear and do not, as written or implemented, provide enough direct incentive for industrial participation. The panel also recognized that in the past, environmental issues have been regarded as a plague by business. Panel participants felt that this must be changed and that a constructive dialogue should be initiated between government and industry. The issue of the environmental impact of decreased phosphate content in detergents underscores this overall need.

About twenty years ago, the eutrophication of lakes aroused increasing concern in the industrialized countries. The source of the problem was thought to be a function of increased use of detergents containing phosphates. Industry adopted a conciliatory approach to government concern. Fifteen years of research were spent trying to find suitable substitutes for phosphates, but no suitable replacement could be found that was as safe. This can be construed as a diversion of resources from other areas of critical need, especially since the polymer bottle which absorbs phosphates has been developed as the container for liquid detergents. This diversion of resources might have been avoided if institutional provisions had been made for a continuing dialogue between government and industry. The need for an appropriate ongoing mechanism for dialogue was reinforced throughout the discussion.

Clear business incentives must exist for industry to participate in environmental programs resulting from international agreements. For example, the high cost of environmental controls has now made it attractive for industry to cooperate in the development of environmental standards and the development of control devices. Particularly with respect to established standards which can reduce uncertainties associated with a multitude of projects, this cooperation can result in more accurate cost-benefit analyses, a reduction in risks of delays, and optimized capital investments. However, international licensing agreements have, at least in one instance, erected barriers to cooperation rather than removed impediments as expected. Those barriers resulted from lack of integration between scientific and economic goals.

One obvious pitfall to avoid in the implementation of these agreements is government pressure for industry to participate when proper business incentives do not exist. Noranda participated in an agreement between Canada and the OECD concerning the production of hydrogen by advanced electrolytical techniques, but then had to terminate participation because of proprietary concerns.

Most of the international agreements do establish the arena for exchange of technical data, and many make provisions for exchange of specialists and for visits of scientists among countries. It was unclear, in the panel's deliberations, under what circumstances industry would be willing to participate in research task sharing or cost sharing. But there was a sense that in areas critical to developing new markets in underdeveloped countries, industry might be willing to share the expense and tasks of gathering environmental information because it would be valuable for long-range strategies.

New business opportunities can also exist in the design of new processes and products keyed to environmental objectives. In fact, there have been examples where environmental regulation has provided opportunities for new businesses. When paints containing chromium were banned, a whole new series of molybdenum pigments were developed.

In recent years, the scientific conference has been used as a model of the United Nations and its family of agencies for the organization of intergovernmental debates on the environment and related subjects. These conferences have resulted in the redirection of research funding and priorities and in newly discovered interests and challenges. Participation of the industrial sector in the conferences could serve to help clarify issues related to the concerns of the industrial sector and could direct funding toward areas where the relationship between productivity and environmental control is critical. It was not clear to the panel whether funds for R&D were made available to implement these agreements, or if so, under what conditions.

The uncertainty about access to funds and participation in the implementation of many of these agreements may result from the circumstance that we are still in an experimental phase of adapting existing international institutions or fashioning new ones to address such problems. There are currently many international institutions, both intergovernmental organizations and nongovernmental entities, devoted to environmental issues which have conducted or participated in extensive cooperative environmental research programs.

PROGRAMS OF INTERNATIONAL ORGANIZATIONS

A review of the activities of international institutions such as those below indicates that their objectives are consistent with or derived from multilateral or bilateral environmental agreements. They may very well provide a suitable structure for the participation of the industrial sector in the implementation or design of these agreements.

The United Nations Environmental Program (UNEP) sponsors a worldwide environmental program linked to social and economic goals and participates in several multi-lateral agreements.

The World Meteorological Organization (WMO) has formulated a world climate program which will consider the role of carbon dioxide in the climate system and assess its ultimate impact on human activities.

The Organization for Economic Cooperation and Development (OECD) is active in collecting and disseminating economic and environmental information and channeling these resources to developing countries.

The Commission of European Communities (CEC) funds research in four general areas: the establishment of criteria for pollutants and environmental contaminants research; environmental information management; research on reduction and prevention of pollution; and research concerning the protection of the natural environment.

The International Atomic Energy Agency is devoted to the solution of problems such as the disposal of nuclear waste and the transport of hazardous nuclear materials.

The International Institute for Applied Systems Analysis (IIASA) is dedicated to exploring new ways to understand and manage complex systems in the areas of environment, energy, and water, among other things. In 1979 this institute had a budget of over $9.6 million.

The International Council of Scientific Unions' (ICSU) special scientific committees have contributed greatly to the body of environmental knowledge. The ICSU's International Biological Program (IBP) coordinates research on the biological basis of productivity, and human adaptability to environmental changes. Ecosystems of major economic importance such as forests, grazing land, and arid lands have been studied. One committee of the ICSU has addressed itself to the needs for interdisciplinary approaches to environmental problems. This committee, the Scientific Committee on Problems of the Environment (SCOPE) has two main tasks: to advance knowledge about the influence of human activities in the environment and to serve as a nongovernmental source of advice in environmental problems. Many international agreements, especially those addressed to bilateral or regional problems, provide for their own institutions for implementation of the agreements.

The International Joint Commission (IJC) on the Great Lakes is a bilateral board with representatives from the United States and Canada and is a mechanism for cooperation between the two countries on matters related to the Great Lakes. It has identified persistent major water quality problems in the Great Lakes: toxic chemicals, high phosphorous additions, contributions from airborne pollutants, and disposal of municipal and hazardous industrial wastes. The commission is currently making an inventory of toxic substances discharged into the Great Lakes, which will be completed in 1982. The IJC's responsibility in the Great Lakes stemmed from the 1972 Great Lakes Water Quality Agreement, and its existence was continued when the agreement was revised in 1978.

Other commissions include the Joint Environmental Commission which was formed to recommend to the United States and Panama measures to avoid or mitigate any adverse environmental effect related to implementation of the new Panama Canal treaties. In addition, there are several other commissions which monitor specific environmental aspects of international agreements. The International Boundary and Water Commission has tried to solve urgent border sanitation problems and is also seeking permanent solutions to specific problems. The International Whaling Commission is providing a framework for the management of whales to ensure their survival.

There are also several international or internationally focused institutions concerned with fostering technology transfer and research cooperation between developed and developing regions. Many of these sponsor programs in the area of environmental or health sciences. These include the International Development Research Centre (IDRC) established by the Parliament of Canada in 1970, the Swedish Agency for Research Cooperation with Developing Countries (SARCDC), the International Foundation for Science (IFS), which gives grants to young scientists for research relevant to development needs, and the prospective United States Institute for Scientific and Technological Cooperation (ISTC). The ISTC is designed to be a grant-making as well as coordinating body and could very well benefit from the expertise of the private sector.

OPPORTUNITIES FOR PRIVATE SECTOR PARTICIPATION

While most of the activities sponsored by these international institutions are carried out by academic and governmental institutions, their activities would certainly be made stronger through the participation of the private sector. Although participation in these activities by individuals from the private sector may be widespread, it seems that an opportunity exists for developing a mechanism to facilitate private sector participation institutionally, especially since many of the research concerns are linked to social and economic issues.   Such institutional participation of the private sector could strengthen the global infrastructure of international institutions, which should in turn strengthen the international resources to ensure sound agreements which can be successfully implemented.

Although no particular mechanism for integrating the private sector's participation in international scientific cooperation in the environmental science field has been defined by the panel, the sense of the panel was that industry could advise and help clarify issues of priority to productivity. The panelists were particularly anxious to have realistic cost-benefit analyses made integral to the design of these programs. Perhaps an international, industrial, environmental forum might address such issues, or be a resource from which to draw. Any mechanisms which would increase the dialogue between government and industry in this area should be of help. One vehicle suggested to facilitate this dialogue was university coordination.

The need for international cooperation in the environmental area is clear.   Moreover, the need to ensure environmental quality, coupled with the need to optimize the efficient use of productive resources, is creating a new set of incentives for partnership involving the public and private sectors of the industrialized countries. Mechanisms which provide opportunities for ongoing constructive dialogue to achieve desired social and economic objectives are the first priority in developing these partnerships.

List of Conference Panels

Panel Members: Kathryn Arnow, William J. Arrol, Gillis Een, Pierre Feintuch, Wolfgang Finke, P.M. Fourt, Gösta Lagermalm, William A. Ragan, Ciril L. Silver, Hugo Thiemann

Rapporteur: Lois S. Peters

ELECTRONICS

Panel Chair: E.F. de Haan

Panel Members: David Beckler, Lee L. Davenport, Claude Dugas, Jacob Goldman, J.V. Harrington, Hiroshi Inose, Ursula Kruse-Vaucienne, Luigi Mercurio, Henry J. Novy, A.E. Pannenborg, Rolf Piekarz, Victor Ragosine, Roland Schmitt, Cyril Silver, K.H. Standke, George Stuart, Michiyuki Uenohara

Rapporteur: Richard Langlois

ENERGY CONVERSION AND CONSERVATION

Panel Chair: Lionel Boulet

Panel Members: Edward L. Brady, Jean Cantacuzene, Gillis Een, Pierre Feintuch, Heinz Gerrens, A.J. Hale, Philip W. Hemily, R. Kahsnitz, Gösta Lagermalm, Rudolf Meier, Yann Moulier, Hans Jürgen Rosenkranz, Reinhard Schulz, Lucien Slama, P.M. Sorgo, and H.R.J. Waddington

Rapporteur: Judith Ugelow

TELECOMMUNICATIONS

Panel Chair: Lee L. Davenport

Panel Members: David Beckler, M.V. Bodnarescu, Claude Dugas, J.V. Harrington, Hiroshi Inose, Ursula Kruse-Vaucienne, Luigi Mercurio, Henry J. Novy, A.E. Pannenborg, Maurice Papo, Rolf Piekarz, R.W. Schmitt, Cyril Silver, K.H. Standke, George Stuart, and Michiyuki Uenohara

Rapporteur: Richard Langlois

TRANSPORTATION

Panel Chair:	Bernhard Schmidt
Panel Members:	William J. Arrol, Wolfgang Finke, Herbert I. Fusfeld, J.E. Goldman, R. Kahsnitz, Eugene G. Kovach, Pekka Rautala, Volker Weiss, Francis Wolek, and H.R. Wynne-Edwards
Rapporteur:	Monica Kaufmann

ENVIRONMENTAL CONTROL

Panel Chair:	Peder Waern-Bugge
Panel Members:	Kathryn Arnow, Duncan Davies, Jorgen Fakstorp, Robert G. Hawkins, Thomas M. McCarthy, Hans J. Rosenkranz, John D. Sheehan
Rapporteur:	Lois S. Peters

List of Conference Participants

CONFERENCE ON INDUSTRIAL PRODUCTIVITY
AND INTERNATIONAL TECHNICAL COOPERATION
November 20-21, 1980
Paris, France

Pierre Aigrain
Secrétaire d'État à la Recherche
France

Dieter Altenpohl
Vice President Technology, Zurich
Swiss Aluminum Ltd.
Switzerland

Kathryn Arnow
Adviser, Scientific Affairs Division
NATO
Belgium

William J. Arrol
Research Management Consultant
W.J. Arrol Consultants
United Kingdom

David Z. Beckler
The Director for Science, Technology & Industry
O.E.C.D.
France

Gerhard Bier
Mitglied des Vorstandes
Dynamit Nobel AG
Germany

M.V. Bodnarescu
Abteilungsdirektor, Leiter der Stabsabteilung
Fried. Krupp GmbH
Germany

Lionel Boulet
Director of Institut de recherche de
l'Hydro Québec
Canada

Edward L. Brady
Associate Director for International Affairs
National Bureau of Standards
United States

Johan Brink
Scientific Counsellor
South African Embassy
France

K.H. Büchel
Member, Board of Management
Bayer AG
Germany

Pierre Burnier
CEM-CERCEM
France

Francis Cambou, Director
Conservatoire National des Arts et Métiers
France

Jean Cantacuzène
Professor of Chemistry, University of Paris
VI and former Scientific Counselor of the
French Embassy in the United States
France

J. Hoyt Chaloud
Manager, Research and Development
Coordination-International
Procter & Gamble
United States

P. Clément
Directeur des Recherches
Administrateur de Kodak-Pathé Vincennes
Kodak-Pathé
France

Charles Crussard
Scientific Director (retired)
Pechiney Ugine Kuhlmann
France

Lee L. Davenport
Vice President & Chief Scientist
General Telephone & Electronics Corp.
United States

Duncan Davies
Chief Scientist and Engineer
Department of Industry
United Kingdom

François Davoine
Le Conseiller technique auprès du Directeur
Conservatoire National des Arts et Métiers
France

Bernard Delapalme
Directeur de la Recherche Scientifique et Technique
Elf Aquitaine
France

Jacques Desazars de Montgailhard
Administrateur Directeur Général
Pechiney Ugine Kuhlmann
France

Claude Dugas
Scientific Director
Thomson-CSF
France

Gillis Een
Director, Strategic Process Planning
Alfa-Laval AB
Sweden

Jorgen Fakstorp
Vice President, Director of
 Industrial Equipment Division

F.L. Smidth & Co.
Denmark

Pierre Feintuch
Controleur Général
Électricité de France
France

Wolfgang Finke
Ministerialdirektor
Bundesminister für Forschung
 und Technologie
Germany

P.M. Fourt
LeDirecteur Général Adjoint
Creusot-Loire
France

C. Fréjacques, Directeur
Delegation Générale à la Recherche
 Scientifique et Technique
France

J. Friedel
Professeur, Laboratoire de Physique des Solides
Université Paris-Sud
France

Herbert I. Fusfeld
Director, Center for Science and
 Technology Policy
Graduate School of Public Administration*
New York University
United States

W.H. Gauvin
Director of Research & Development
Noranda Mines Ltd.
Canada

Heinz Gerrens
Director of Process Development,
 Corporate Research & Development
BASF Aktiengesellschaft
Germany

---

*The Center for Science and Technology Policy became part of the
 Graduate School of Business Administration in September 1981.

J.E. Goldman
Senior Vice President & Chief Scientist
Xerox Corporation
United States

E.F. de Haan
Senior Managing Director of Research
N.V. Philips Gloeilampenfabrieken
The Netherlands

Carmela S. Haklisch
Assistant Director, Center for
 Science and Technology Policy
Graduate School of Public Administration*
New York University
United States

Arthur J. Hale
Vice President, Preclinical R&D,
 European Operations
G.D. Searle & Co. Ltd.
United Kingdom

J.V. Harrington
Senior Vice President, Research
 and Development and Director,
 Communications Satellite Corp.
United States

Robert G. Hawkins
Vice Dean, Faculty of Business Administration
 and Vice Dean, Graduate School of Business
 Administration
New York University
United States

Philip W. Hemily
Deputy Assistant Secretary General for
 Scientific Affairs
NATO
Belgium

George Hudelson
Vice President-Engineering,
 Director, Research Division
Carrier Corp.
United States

*The Center for Science and Technology Policy became part of the
 Graduate School of Business Administration in September 1981.

Hiroshi Inose
Director of Computer Center and
 Professor of Electronic Engineering
University of Tokyo
Japan

Roland Kahsnitz
Manager of Research
ESSO Europe Inc.
United Kingdom

Monica Kaufmann
Rapporteur
New York University
United States

Eugene G. Kovach
Deputy Director,
 Office of Advanced Technology
Department of State
United States

Ursula Kruse-Vaucienne
Technology Planning Specialist
Honeywell, Inc.
United States

Gösta Lagermalm
 Senior Policy Advisor
 Swedish Board for Technical Development
Sweden

Bernard Langley
Member of the Policy Group
ICI Corp.
United Kingdom

Richard N. Langlois
Research Associate,
 Center for Science and Technology Policy
 and member of Faculty, Dept. of Economics
New York University
United States

M. Lavalou
Directeur Général des Recherches
 et du Développment
Rhone-Poulenc S.A.
France

Thomas M. McCarthy
Manager, Professional and Regulatory Relations
Procter & Gamble Europe
Belgium

Rudolf W. Meier
Deputy, Corporate Research
BBC Brown, Boveri & Co. Ltd.
Switzerland

L. Mercurio
Olivetti
Italy

M. Mouflard
Scientific Adviser
Renault Régie Nationale
France

Yann Moulier
Membre du Centre d'Analyse et de Prévision
Ministère des Affairs Étrangeres
France

Henry J. Novy
Novy Eddison & Partners
United Kingdom

A.E. Pannenborg
Vice President, Board of Management
N. V. Philips Gloeilampenfabrieken
The Netherlands

Maurice Papo
Director of Science and Industry Programs
IBM
France

K.J. Parker
Chief Scientist
Tate & Lyle Ltd.
United Kindom

Lois S. Peters
Research Scientist, Center for
 Science and Technology Policy
New York University
United States

Rolf Piekarz
Group Leader, Directorate for Scientific,
 Technological and International Affairs
National Science Foundation
United States

Paolo della Porta
Vice President and Managing Director
SAES Getters
Italy

William A. Ragan
Vice President, Research & Development
Becton-Dickinson
United States

Victor E. Ragosine
Vice President, Advanced Technology
Ampex Corp.
United States

Pekka Rautala
Director of Physics Laboratory, Tapiola
Outokumpu OY
Finland

Hans Jürgen Rosenkranz
Head of Coordination R&D
Bayer AG
Germany

Jöel de Rosnay
Directeur des Applications de la Recherche
Pasteur Institute
France

R. Saint-Paul
Le Président, Département Économie et Gestion
Conservatoire National des Arts et Métiers
France

Jean-Jacques Salomon
Head, Science Policy Division
O.E.C.D.
France

Lewis H. Sarett
Senior Vice President for Science & Technology
Merck & Co.
USA

Bernhard Schmidt
Vice President and Deputy Chairman of
 the Board of Executives
Dornier GmbH
Germany

Roland W. Schmitt
Vice President, Corporate Research and Development
General Electric Company
United States

Reinhard Schulz
Secretary General
E.I.R.M.A.
France

John D. Sheehan
Manager of Licensing
Stauffer Chemical Co.
United States

Cyril L. Silver
Directorate-General for Research,
 Science and Education
Commission of the European Communities
Belgium

Lucien Slama, Directeur
Compagnie Électro-Méchanique
France

Poznan M. Sorgo
Manager, Licensing Technology
N.V. Phillips Petroleum Chemicals SA
Belgium

Klaus-Heinrich Standke
Deputy Assistant Director General
UNESCO
France

Jacques Sterlini
CEM-CERCEM
France

Christian Stoffäes
Chef du Centre d'Études et de Prévision
Ministère de l'Industrie
France

George Stuart
Institut für Systemtechnik und Innovationsforschung
Germany

Eliezer Tal
Institut für Systemtechnik und Innovationsforschung
Germany

Hugo Thiemann
Member of Management Committee
Nestlé SA
Switzerland

Michel Turpin
Directeur Adjoint
CERCHAR
France

Michiyuki Uenohara
Senior Vice President,
 Research & Development & Director
Nippon Electric Co. Ltd.
Japan

Judith L. Ugelow
Rapporteur
New York University
United States

H.R.J. Waddington
Director, Administrative and Scientific Services
Beecham Pharmaceuticals
United Kingdom

Peder Waern-Bugge
Director, Research and Development
Stora Kopparberg-Bergvik
Sweden

Volker Weiss
Vice President for Research and
 Graduate Affairs
Syracuse University
United States

Francis Wolek
Deputy Assistant Secretary for Productivity,
 Technology and Innovation
Department of Commerce
United States

Hugh R. Wynne-Edwards
Vice President and Chief Scientific Officer
Alcan International Ltd.
Canada

Ake Zachrison
Director of Technological Development
AB Volvo
Sweden

Index

About the Editors and Contributors

LIONEL BOULET is Director of the Institut de recherche de l'Hydro-Québec, Canada.

KARL HEINZ BÜCHEL is a member of the Board of Management, Bayer AG, Germany.

JEAN CANTACUZÈNE is a Professor of Chemistry at the University of Paris VI and former Scientific Counselor of the French Embassy in the United States

CHARLES CRUSSARD is retired Director, Scientific Pechiney Ugine Kuhlmann, France.

LEE L. DAVENPORT is Vice President and Chief Scientist, General Telephone and Electronics Corporation, United States.

JACQUES DESAZARS DE MONTGAILHARD is Administrateur Directeur Général, Pechiney Ugine Kuhlmann, France.

HERBERT I. FUSFELD is Director of the Center for Science and Technology Policy, Graduate School of Public Administration, New York University.

W.H. GAUVIN is Director of Research and Development, Noranda Mines, Ltd., Canada.

J.E. GOLDMAN is Senior Vice President and Chief Scientist, Xerox Corporation, United States.

E.F. DE HAAN is Senior Managing Director of Research, N.V. Philips Company, The Netherlands.

CARMELA S. HAKLISCH is Assistant Director of the Center for Science and Technology Policy, Graduate School of Public Administration, New York University

ROBERT G. HAWKINS is Vice Dean, Faculty of Business Administration, and Vice Dean, Graduate School of Business Administration, New York University, United States.

154

BERNARD W. LANGLEY is a member of the Policy Group, ICI Corporate Laboratory, United Kingdom.

RICHARD N. LANGLOIS is a Research Associate at the Center for Science and Technology Policy, Graduate School of Public Administration, New York University and is on the faculty of New York University's Department of Economics.

RUDOLF W. MEIER is Deputy, Corporate Research, of Brown, Boveri & Co., Ltd., Switzerland.

MAURICE PAPO is the Director of Science and Industry Programs for IBM France.

LOIS S. PETERS is a Research Scientist at the Center for Science and Technology Policy, Graduate School of Public Administration, New York University.

LEWIS H. SARETT is Senior Vice President for Science and Technology, Merck & Company, United States.

BERNHARD SCHMIDT is Vice President and Deputy Chairman of the Board of Executives, Dornier GmbH, Germany.

ROLAND W. SCHMITT is Vice President, Corporate Research and Development, General Electric Company, United States.

KLAUS-HEINRICH STANDKE is a Principal Director with the United Nations Educational, Scientific and Cultural Organization (UNESCO).

MICHIYUKI UENOHARA is Senior Vice President, Research and Development, and Director of Nippon Electric Company, Ltd., Japan.

PEDER WAERN-BUGGE is Director, Research and Development, Stora-Kopparberg-Bergvik, Sweden.

AKE ZACHRISON is Director of Technological Development, AB Volvo, Sweden.